A Handbook on Post Harvest Management of Fruits and Vegetables

THE AUTHOR

Dr. P. Jacob John, presently working as Professor Head of the Department of Processing Technology, College of Horticulture, Kerala Agricultural University, Vellanikkara, Thrissur, Kerala, is an Agricultural graduate from Kerala Agricultural University and later took his M.Sc and PhD in Food Technology from CFTRI Mysore. He has nearly 30 years of experience in teaching, research and extension activities ub the university. He has produced more than 30 postgraduates from his department as HOD and personally guided nearly 10 PG students in post harvest technology. He is a member of the board of studies of Calicut University, Chairman, board of studies of Mahatma Gandhi University and Programme incharge of IGNOU study centre for diploma course in Fruit and vegetable processing. He is the co-ordinator of post harvest technology research programme of the University. He was the president of the Association of Food Scientists and Technologists (AFST India), for the year 2003-04, member in the ICAR accreditation team for restructuring the PG courses in Food Science and Technology on State Agricultural Universities during 1999. He has in his credit several national and international papers, popular articles and patents in the field of specialization.

A Handbook on Post Harvest Management of Fruits and Vegetables

Dr. P. Jacob Johan

2024

Daya Publishing House®

A Division of

Astral International Pvt. Ltd.

New Delhi – 110 002

ISBN 9788170355328 (HB)

Published by : **Daya Publishing House**®
A Division of
Astral International Pvt. Ltd.
– ISO 9001:2015 Certified Company –
4736/23, Ansari Road, Darya Ganj,
New Delhi-110 002
Ph. 011-43549197, 23278134
E-mail: info@astralint.com
Website: www.astralint.com

Digitally Printed at : **Replika Press Pvt. Ltd.**

Foreword

India is a global leader in production of fruits and vegetables. We share 12 percent of world's production in fruits and 15 percent in case of vegetables. India is the second largest producer of fruit, next to Brazil and vegetables, next to China in the world. In general productivity of fruits has nearly doubled from 5.5 to 10.3 t/ha; over the years. In spite of these impressive productivity records, our over all performance in this field is far from satisfactory due to an estimated annual post-harvest handling loss of 25 - 30 %. These losses are mainly due to lack of processing facilities, knowledge of post-harvest practices, infrastructure facilities to handle perishables products like fruits and vegetables and also due to inadequate packaging, transport and marketing strategy.

So there is an urgent need to emphasize on this important aspect. Considering this, it is now taken up as one of the important subjects for the undergraduate and post graduate students in Agriculture and Horticulture. In order to fulfill this requirement an attempt has been made to compile and present important aspects involved in the protocol for the proper post-harvest management of fruits and vegetables, like harvesting, sorting, grading, packaging, pre-treatments and storage techniques in to a small hand book, titled **"A Hand book on post harvest management of fruits and vegetables".** I am sure that this publication will lead to more scientific and methodological approach to post harvest handling of highly perishable commodities like fruits and vegetables. This approach can considerably reduce the post-harvest losses, add value to these commodities and increase the profit of farmers. This will also generate employment and entrepreneurship opportunities for the rural youth.

I hope this will serve as a ready-reckoner for packers, exporters, students and researchers in the field of fruits and vegetables.

K.R. Viswambaran, ***IAS***

Vice-Chancellor

Kerala Agricultural University

Preface

India is considered to be the second largest producer of fruits and vegetables in the world. The production of fruits and vegetables will have significance only when they reach the consumer in an acceptable condition and at a reasonable price.

Unfortunately, in spite of higher production achieved in the field of horticultural crops, a considerable gap exists between gross production and net availability of fruits and vegetables due to huge post harvest loss. These losses are again due to lack of knowledge in the proper post harvest management or handling of fruits and vegetables right from harvesting to marketing. It is therefore imperative that post harvest management of horticultural produce is essential to feed those engaged in production, research and teaching in post harvest aspects.

Considering this aspect it is now taken up as one of the important subject for the undergraduate and postgraduate students of Agriculture and Horticulture. Even though informations on harvesting, washing, sorting, grading post harvest treatments, packaging, storage and transportation are available elsewhere in different books and websites, not a single, handy book exclusively for this topic is not so common; there fore it was thought to publish my lecture notes prepared for the UG students into a small hand book compiling the informations gathered from different sources.

I presume that, this will be of much use to those who are engaged in teaching, research and learning in post harvest aspects of fruits and vegetables as a ready reckoner.

P. Jacob John

Contents

Chapter 1

Introduction

India is the one and only one country blessed by 'Mother Nature' to have the production and availability of one or more fruits at any point of the year. With a production increase of over 300 per cent during the last three decades, India today stands second in the world production of fruits and vegetables. Yet, India's share in world trade is miniscule that is less than one per cent.

While area and production have been increasing at a fast pace, the present infrastructure of post harvest management is very poor. Major constraints of Indian fruits and vegetables are heavy post harvest losses in the range of about 20-30 per cent.

Harvesting at proper maturity, adoption of improved harvesting and handling techniques have to be looked into as priority. Innovations in mechanical harvesting, electronic sorting, proper handling, grading, packaging, storage and transportation are some of the points to be approached systematically to reduce the post harvest loss. Recent developments in the storage techniques like modified and sub–atmosphere storage, computerized sorting and grading and adoption of cool chain are noteworthy. A lot more need to be done in grading and standardization of storage systems and proper transportation. There need to frame grade specification for all the fruits and vegetables. Palletisation and containerization will go a long way in establishing national and international trade on a firm footing. Therefore proper understanding of all these factors at various levels by the persons involved in handling these commodities is a pre requisite; with this view the present handbook is published.

Agro climatic suitability coupled with abundance of natural resource endowment, equips India with a unique comparative edge in the cultivation of a variety of horticultural crops. In realization of their potential for providing nutritional security, employment generation and foreign exchange earnings, horticulture has been accorded the top priority in the last two year plans.

Among the horticultural commodities, fruits and vegetables owe its own role, contributing over 24 per cent of country's agricultural GDP, more than 10 per cent of the total export earnings and supports more than 19 per cent of the labour force. Fruits and vegetables have its own vital position because of its richness in energy, vitamins, and minerals making it imperative to incorporate fruits and vegetables in to our daily diet. India can be regarded as 'Horticultural Paradise'. India has emerged as the second largest producer of fruits and vegetables in the world.

A wide gap between production and availability exists. Exports constitute only to a small proportion of total production due to the increasing domestic demand, consumer preference, varietal characters, quality, domestic and international prices and inadequate infrastructure facilities for post harvest handling methods.

Table 1: Area and Production of Fruits and Vegetables (2004–05)

Item	*Area (million ha)*	*Production (million tonnes)*	*Position in World*
Fruits	4.66	49.30	II
Vegetables	67.56	101.43	II

Table 2: Availability and Requirement of Fruits and Vegetables

Item	*Per Capita Requirement (g/day/person)*	*Per Capita Availability (g/day/person)*
Fruits	120	99
Vegetables	285	210

Quality assurance is a pre-requisite for high value fresh fruits and vegetables. Proper handling ensures quality, which is preserved until the produce reaches the consumer.

Wide gap between production and availability of fruits and vegetables exist mainly due to loss. About 20 to 30 per cent production of fruits and vegetable is lost after harvest.

"Loss is defined as any change in the availability, wholesomeness or quality of food".

Post Harvest Loss

All those losses that occur between the period of separation of healthy commodity from the parent plant till it reaches the consumer.

Post harvest loss of fresh fruits and vegetables are estimated to be 20-30 per cent and they are classified mainly into 3 types.

Physical Loss

Fruits and vegetables have very soft texture, high moisture content and are very much susceptible to mechanical injury. Poor handling, unsuitable containers, improper packaging and transportation can easily cause bruising, distortion and other forms of injury.

Physiological Loss

Fruits and vegetables are still alive after harvest and continue their physiological activity. Rate of respiration, gaseous exchange, transpiration etc. are hastened. This occur mainly due to mineral deficiency, low or high temperature injury or undesirable atmospheric conditions such as high humidity.

Biological Loss

By invasion of fungi, bacteria, insects and other organisms, control of post harvest decay is becoming a more difficult task because the number of pesticides available for the control is falling rapidly as consumer concern for food safety increases.

The grave situation of post harvest loss can be realized from the findings of Food Ministry of India that we waste more fruits and vegetables than that the U.K. consumes every year. Post harvest loss can be considered as a social evil, which eats up the grower's margin and pushes up the consumer's price. This post harvest loss is mainly due to improper and unscientific handling and packing being followed in India.

Post harvest loss of perishables may be due to physical, physiological or pathological reasons or a combination of these factors. Following table shows the principal causes for the loss of major group of fruits and vegetables.

Table 3

Group	*Commodity*	*Principal Causes for Loss and Poor Quality*
Root vegetables	Carrot	Mechanical injury
	Beetroot	Improper handling
	Onion	Sprouting and rooting
	Sweet potato	Water loss (shriveling)
	Garlic	Decay
Leafy vegetables	Lettuce	Water loss (wilting)
	Spinach	Loss of green colour
	Cabbage	Mechanical injury
		Decay
Flower vegetables	Broccoli	Mechanical injury
		Yellowing and other discolourations
	Cauliflower	Abscission
		Decay
Immature vegetables	Brinjal	Over maturity at harvest
	Okra	Water loss
	Cucumber	Bruising and other mechanical injuries
	Squash	Decay
Mature fruits and	Tomato	Bruising
vegetables and all fruits	Melons	Over ripeness and excessive softness at harvest
	Citrus	Water loss
	Banana	Compositional changes
	Mango	Decay
	Apple	

In a developing country like India because of the poor post harvest management, estimated loss of 20-30 per cent is often met in horticultural produce which amounts in crores per annum. This results in widening the gap between the demand and supply making the per capita consumption of fruits and vegetables far below the recommended quantity. Therefore in order to minimize this loss it is essential to understand and control various factors that contribute to the loss at every step right from harvesting till it reaches the consumer.

The magnitude of these losses is going to increase with more area expansion and production if proper post harvest interventions are not engineered.

An increase in the rate of loss because of normal physiological changes is caused by conditions that increase the rate of natural deterioration such as high temperature, low atmosphere humidity and physical injury. Abnormal physiological deterioration occurs when fresh produce is subjected to extremes of temperature. This may cause unpalatable flavours, failure to ripen or other changes in the living processes of the produce, making it unfit for use.

Careless handling of fresh produce causes internal bruising, which results in abnormal physiological damage or splitting and skin breaks, thus rapidly increasing water loss and the rate of normal physiological breakdown. Skin breaks also provide site for infection by disease organisms causing decay.

All living material is subjected to attack by parasites. Fresh produce can become infected before or after harvest by diseases widespread in the air, soil and water. Some diseases causing organisms are able to penetrate the unbroken skin of produce while others require an injury in order to cause infection. Damage so produced is probably the major cause of loss of fresh produce.

The influence of all these causes are strongly affected by various stages of post harvest operations. Furthermore, they all have great effect on the marketability of the produce and the price paid for it.

Fruits and vegetables both fresh or ripe, can be a valuable commodity both as food for the people and as a cash crop for its producer. Though utmost care is taken till its harvest, less care and importance is often given to quantity as well as to the quality.

Harvesting at proper maturity, careful harvesting techniques, proper handling, reducing the field heat, adequate transportation etc. can reduce losses to a great extent.

Horticultural crops pose more storage problems than any other food stuff because they have very high moisture content and they are physiologically more active. The two basic processes, respiration and transpiration, cause some loss in food values during storage. Besides secondary causes such as physical damage and microbial spoilage are also responsible for much higher losses. Although the exact information on losses is not available, it is estimated that about 25 to 30 per cent of the produce goes to waste on account of inadequate post harvest management, technologies and infrastructural facilities. No nation can afford to continue with such a depressing and disappointing situation. It would be, therefore, futile to expect progress of horticultural industry without strong support of the post harvest handling and management technologies.

Figure 1: Steps Involved in Post Harvest Management

Proper harvesting

↓

Proper Washing

↓

Sorting and grading

↓

Pre –treatments

↓

Packaging

↓

Storage

↓

Transportation

References

1. Chadha, K.L. 2000. An overview of Hi-Tech Horticulture: Opportunities and constraints. Proceedings of National seminar on Hi-Tech Horticulture, Bangalore, June 26, pp.13.
2. FAO 1985. *Prevention of Post-Harvest Food Losses: A* Training Manual. Rome: UN FAO. P.120.
3. Madan, H.S. and Ullasa, B.A. 1993. Post harvest losses in fruits. Advances in Horticulture. (Ed.) Chadha, K.L, and Pareek, O.P. Malhotra Publishing House, New Delhi. 4(4) p: 1795.

4. Pruthi, J.S. 1993. Innovations in post harvest technology of vegetables. Advances in Horticulture. (Eds.) Chadha, K.L. and Pareek, O.P. Malhotra Publishing House, New Delhi. 6(2) : 1153.

5. Thompson, A.K. 1996. Post harvest technology of fruits and vegetables. Blackwell Science Ltd. pp.147-169.

6. Uppal, D.K., Sharma, S.K. and Sharma, R.K. 1994. Post harvest management of horticultural produce challenges before India. Paper presented in the Seminar on Management of Post harvest technology in horticulture, 23 May, 1994, National Institute of Rural Development.

Chapter 2

Harvesting

Method and time of harvesting and the care taken during harvesting are important. Considerable harm to the fruits or vegetables is caused by careless harvesting or exposing the harvested produce to excessive field heat, if it is not harvested at the proper time. Harvesting is to be done during the early hours of the day or late in the evening. Temperature above 27°C during harvesting should be avoided. There should not be any dew on the commodity nor should it be wet while harvesting. Harvested commodity is to be heaped under shade until it is packed. Harvesting at optimum maturity ensures better cold storage life and better sensory quality.

Indices for judging this maturity stage of different fruits and vegetables have been worked out and the standards laid down in most cases. But the growers and more often the pre-harvest contractors resort to harvesting immature fruits to catch the early market.

Another problem related to maturity is the mixing up of fruits of different maturities in the same pack. Tomatoes ranging from red ripe to green stages are seen packed together in a basket or box. While this system satisfies the retailer who needs ripe fruit for immediate sale, it causes transportation losses exceeding 30 per cent and the consumer pays heavily for this loss.

Harvesting Time

Harvesting is to be done during the early hours of the day or late in the evening. If it is for near market we can harvest early in the morning whereas for distant market it can be harvested late in the evening and can be transported during night.

Method of Harvest

Damage and impact caused during harvesting process tells upon the shelf life of any produce. Shelf life is reduced by the impact of harvest, therefore suitable harvesting gadgets must be used where ever possible; example telescopic type of poles are used for harvesting mangoes or a person standing on a hydraulic lift for directly collecting the commodity. Bamboo pole with net attached to it can also be used for harvesting mangoes, which can reduce the impact shock while harvesting.

A prototype hand held mango harvester is available. It consists of an adaptation of the commonly used pole with a bag attachment. Mechanical aids to harvesting vegetables such as lettuce, cauliflower and cabbage involve cutting the vegetable by hand and placing it on a conveyer in a mobile packing station, which is slowly conveyed across the field.

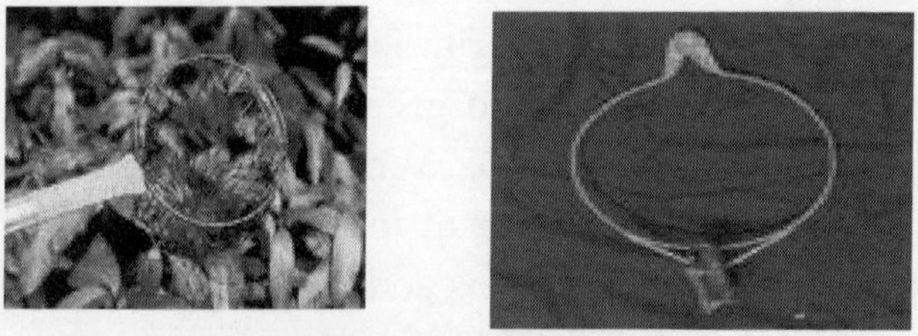

Figure 2: Mango Harvester

Maturity

Depending upon the purpose of the commodities after harvest it can be classified into two maturities.

Physiological Maturity

Stage in the development of fruit and vegetable when maximum growth and maturation has occurred.

Horticultural Maturity

Stage of development when a plant or plant part posses the pre requisite for utilization by consumer for a particular purpose.

Maturity Indices of Fruits and Vegetables

The general limitations of all maturity indices are variations in (a) fruit size (b) position of the fruit in the tree (c) climate and seasonal effects (d) fertilizer and manures (e) soil type (f) soil moisture (g) pruning methods and use of hormones and other chemicals. In spite of all these limitations it is still possible to combine various indices of maturity to assess the stage at which the commodity may be harvested. Some of the examples are as follows.

Avocado

For the proper storage of avocado the stage of maturity at harvest is important. Too immature fruits may be avoided since they tend to be inferior in flavour and texture on subsequent ripening.

Presently, criteria such as size, colour etc may be evaluated on the basis of experience are in practice. No other objective methods are available for avocado. However, for two varieties in California, the state regulation specifies a minimum of 8 per cent oil. Florida has several commercially important varieties; they vary in oil content from 3 to 7 per cent. At present Florida avocado industry is using minimum fruit weights and diameters for each commercial variety in conjunction with the picking dates.

Banana

When the banana are to be transported to distant places, they are picked slightly immature at about 75 to 80 per cent maturity, with plainly visible angles and will ripen in about 3 weeks.

Bananas for inter–island shipment are harvested when about 85 to 90 per cent maturity, when they attain full development, but the fruit angles are still not very well defined. The fruits ripen from 1 to 2 weeks after harvest. For local or near by markets more mature fruits are harvested and they ripen in less than a week.

For judging maturity, pulp to peel ratio, days from the emergence of inflorescence, disappearance of angularity of fingers, drying of the leaves, brittleness of the floral ends are some of the indices used in India. Angularity or fullness of fingers seems to be the standard practice. To determine the proper time of harvesting it is best to supplement "fullness of finger" with size and the number of days it takes from inflorescence emergence to maturity. "Dwarf cavendish" banana usually takes 90 days to reach the "full three quarter" stage of maturity after fruit set. This has pulp to peel ratio of 1.3 to 1.45 and is considered ideal for distant transportation.

Stage of maturity in Robusta variety cannot easily be detected by pulp/ peel ratio; but other tests like number of days from the time of shoot emergence and harvesting and fruit growth measurements are reliable indices of maturity. 135-145 days between shoot emergence and harvesting is equal to 8-9 ring numbers with fruit girth measurement as 4.25 to 4.5 cm.

Grapes

" Ana bee–shahi" or selection –7 are harvested when the average TSS reaches 15 to 16 per cent. A range of 18-20 per cent TSS is considered as a good index for "Thomson Seedless" variety.

'Bangalore Blue' which is mostly used for juice and wine, is harvested when the juice records a TSS of 12 to 14 per cent.

Besides some physical characteristics like texture of the pulp, peel colour, easy separation of the berries from the bunch and development of characteristic flavour and aroma are useful indices.

Jackfruit

A dull hollow sound is produced when the fruit is tapped by the finger, the last leaf of the peduncle turns yellow, the fruit spines become well developed and wide spread, the spines yield to moderate pressure and the typical aroma develops.

Jackfruit intended for immediate consumption is harvested when the rind is fairly soft, the peduncle leaves have turned orange and the fruit produces an aroma. At this stage the flesh is slimy, juicy and orange yellow in colour.

For distant market, it should be harvested when still firm and without any aroma. The leaf nearest to the fruit must be starting to turn yellow and the spines already well developed and flesh turns to pale yellow in colour.

Citrus

Maturity indices differ among citrus varieties. In 'Valencia' and 'Ladu" at least 25 per cent colour break is considered mature. In citrus varieties that do not de green completely, periodic sampling and tasting will indicate fruit maturity ideal for harvesting. In India, Coorg mandarins are harvested from December to January, when the rind colour changes from green to orange, the juice has an acidity of 0.4 per cent and TSS 12 to 14 per cent. Sweet oranges are harvested from September to October when the rind turns to yellow, the acidity of the juice is 0.3 per cent and TSS 12 per cent.

Table 4: International Standards for the Minimum Juice Contents of Citrus Fruits

Thomson naval orange	30 per cent
Other orange varieties	30 per cent
Grape fruit	35 per cent
Lemon	25 per cent
Mandarins	33 per cent

Sweet Oranges

Maturity standards are based on juice yield, TSS and Brix-acid ratio. These vary depending upon the areas of cultivation and season. The characteristics of highest quality include firmness, good weight for size, skin and flesh colour and skin texture.

Musambi (*Citrus sinensis*) grown in Andhra Pradesh is bright orange yellow, while sathugudi (C. *sinensis osbek*) grown in Tamil Nadu is greenish yellow when fully mature. Some oranges during early harvest contain chlorophyll in the peel. This is due to the active growth condition of the tree coinciding with ripening of the fruit, thus additional accumulation of chlorophyll in the peel termed re-greening, which does not in any way affect the eating quality of the fruit.

Mandarin

In Coorg, fruits are harvested from December to January when colour changes from green to orange. The acidity is 0.4 per cent and TSS range from 12 to 14 per cent. The desirable characters of the mandarin group are smooth surface, thin skin, good weight in relation to size, high colour for particular size, freedom from physical damage or decay and minimum juice content to be 33 per cent.

Lemon

Harvested when the surface is still green. The colour develops during storing and prolonged transit. Lemons of good quality must be light to medium yellow, firm, smooth skinned and heavy for their size.

Limes

Yellow to slight orange yellow colour. Fruits must be smooth skinned, free from scars.

Mango

Changes associated with maturity of mangoes are (1) fullness of shoulders (2) changes in the colour of the pedicel (3) growth of stones and (4) development of lenticels.

Several growers depend on the change of peel from deep green to olive green colour. "Alphonso and Pairi cultivars" usually take 110 to 125 days after fruit set to reach optimum maturity, the flesh colour turns from white to pale yellow. In some varieties it is determined by noting degree at which the shoulder extends out of the stem end.

Specific gravity measurement proved to be a reliable index. Specific gravity may be affected by size of the seed cavity, rainfall and cultural practices. Specific gravity between 1.01 and 1.02 are suitable for picking.

Picking date will differ with variety and growing area. For each variety in a particular area it is essential to determine the date from flowering to maturity. It is suggested that starch to acid ratio [4 or more] could be used as an index for maturity in Langra variety. Starch content 5 per cent at the time of harvest could be a reliable index.

Maturity of mangoes grown in India may be designated into 4 stages *viz.* A,B,C and D

Stage–A

The fruits have their shoulder in line with the stem and skin colour green.

Stage–B

The shoulders have out grown the stem end. This is the best stage for export.

Stage–C

The colour has lightened towards yellow.

Stage–D

The fruits are fully ripe with typical colour developed on the skin.

Table 5: Maturity Standards for Harvest of Alphonso and Pairi Mangoes

Physical and Chemical Factors	Maturity Group			
	Over Mature	Physiologically Mature	Physiologically Immature	Physiologically Immature and Under Size
	A	B	C	D
1. Weight (g)	>320	300±20	250±20	<225
2. Specific gravity	>1.02	1.01-1.02	1.0-1.01	<1.0
3. T.S.S (per cent)	>10	8±1	7±1	<6.0
4. Acidity (% as malic acid)	<3.2	3.5±0.2	3.9±0.2	>4.1
5. Total carotenoid (microgram)	>80	600-800	400-800	<400
6. *AIS (per cent)	>12.5	11.5-12.5	11.5-12.5	<10.5

* Alcohol insoluble solids.

Papaya

If intended for local market, they are left on the tree until firm ripe stage. At this stage, a change of colour at the apical end of the fruit occurs. As soon as the trace of yellow appears on the apex or between ridges, the fruit should be removed. This stage of harvest takes 4 to 5 days to ripen. Other indices are TSS should be 6 per cent, and the surface should be one-third coloured. For the development of maximum TSS in ripe fruit after harvest, the fruit should have at least 33 per cent of the surface turned yellow in colour.

Pineapple

The stage at which it is to be harvested depends on the ultimate destination or use. Fruit for home use is picked when 25 per cent yellowing is obtained. At this stage the fruit has higher TSS and low acidity.

If the fruit is intended for long distance markets, usually harvested when all the 'eyes' are still green and have no trace of yellow colour. It takes 2-3 weeks for ripening.

When immature, the 'eyes' are grey or light green and the small bracts which cover one half of each eye are grey or almost white giving a greyish appearance.

When the fruit matures the space between the 'eyes' fill and the colour gradually changes from light to dark green. The following shell colours of pineapple are generally used to determine the various stages of maturity.

No. 0. All eyes are totally green.

No. 1. Not more than 20 per cent of the eyes are yellow.

No. 2. More than 20 per cent and less than 40 per cent of the eyes are tinged with yellow.

No. 3. More than 55 per cent and less than 65 per cent eyes turned yellow.

No. 4. Above 65 per cent and less than 90 per cent eyes are full yellow.

No. 5. More than 90 per cent yellow but not more than 20 per cent of the eyes are reddish orange.

No. 6. 20 to 100 per cent of the eyes are reddish brown.

No. 7. Shell is reddish brown and shows signs of deterioration.

No. 2 and No. 4 are used for canning and for shipment as fresh fruit.

No. 6 and No. 7 are over ripe and have fermented flavour.

No. 0 is mostly used for distant markets that takes 2 to 3 weeks for ripening.

In India, pineapple is harvested when the colour changes from green to greenish yellow. The fruit develops smooth surface around the eyes, the bracts start drying up. A mature pineapple for canning should have TSS of 12 per cent and acidity of 0.5 to 0.6 per cent.

Vegetables

Tomato

Harvesting of tomato depends on the purpose. Three maturity stages are generally recognized– mature green, pink or breaker and the red ripe stage. For long distance transport, it should be harvested at the mature green stage.

. At the breaker stage the blossom end will turn pinkish or reddish in colour. At ripe stage the surface is pink or red.

Fruits for local or nearby markets and canning purpose, should be harvested at breaker or ripe stage. Fruits at the mature green stage the blossom end of the fruit will turn cream in color, the pulp surrounding the seeds is jelly like and the seeds slip away from the knife.

For canning the fruits should be medium large, smooth, of uniform rich red flesh, evenly ripened without green shoulders, and possessing a large proportion of solid flesh of good flavour. Irregularly shaped tomatoes are difficult to peel. For juice making, tomato is to be rich in colour and flavour and juicy rather than pulpy.

Peas

At the time of harvest, the fruit must be uniform in size and maturity, and shell easily. It must remain green after blanching and processing. The pods should be well filled with young and tender peas, changing colour from dark to light green. Firmness increases with maturity. Hence instrumental methods (Tendrometer) have used for measuring maturity of peas. With the progress of maturity and size, sugar content decreases rapidly, while starch and protein increases. Hence, high sugar content is an indication of quality. The following indices are recommended:

Table 6

Purpose	*AIS %**	*Tendrometer reading*
For canning	13-13.4	115
Freezing	11.1-12.9	95-105
Drying	9-11	85-95

* AIS-Alcohol Insoluble Solids.

Peas are also graded by brine floatation after being graded for size and blanched. Fancy quality peas will float in a brine of 1.04 sp. gravity and standard quality peas in a brine of 1.17 sp. gravity. Peas which sink in heavier brine are designated as substandard.

French Beans

Beans reach optimum stage of maturity between 14 and 18 days of setting. At this stage beans should snap readily and the tip should be pliable. When the seeds have developed and the constrictions are seen, it means that the optimum maturity stage has passed. When the beans become mature or over mature, the cells in pods become lignified and compressed, and numerous fast fibers appear resulting in fibrous texture and unpleasant taste. Hence beans must be harvested at the optimum stage of maturity.

Total solids, crude fibre, alcohol insoluble solids (AIS), maximum shear force and weight to length ratio are used as indices for determination of maturity. For acceptable tenderness the value of various indices should be total solids 8.3 per cent, AIS 4.7 per cent, crude fibre 3.8 per cent, shear force 9.0 kg and weight to length ratio 0.5 g/cm.

Desirable characteristics for French beans used for canning or freezing are long, straight tender pods of medium size which are thick having small seeds, and free from fibre and strings at optimum maturity. During blanching and cooking the beans should retain the bright green colour and should not split.

Carrot

The desirable characteristics of carrot are bright orange colour, absence of fibrous core, tender, crisp and sweet flavour. For canning, the carrot should have a smooth surface and to be free of furrows or wrinkles. The core should not be tough and hardy; and should have good colour similar to flesh. When canned as a whole, the shoulder diameter should be 0.5 and 1.0 inch which is achieved by closer spacing of plants in the field. For drying, carrot must be deep orange colour and be free from woody fibre. For freezing, tenderness is most important. When intended for use in soups, the firm texture should be retained during canning process.

For freezing and canning, it should be harvested while still tender and for dehydration, when slightly mature.

Okra

The fruits are harvested when the pods are still young, tender and exhibiting maximum growth rate. At this stage, the pods readily snap when picked. Mature pods are fibrous, tough and unfit for human consumption.

Okra for canning or freezing, should be medium length, tender, straight with blunt ends and bright green in colour.

Cabbage

Cabbage mature between 62 and 110 days from field setting at low elevation and 81 to 125 days at high elevations. Solidity and firmness are the useful maturity standards used. Colour also is used as an index. The head turns a light shade of green at full development for making "sauerkraut", the head is allowed to reach full maturity, as indicated by curling back of leaves and exposure of cover leaves beneath.

Cauliflower

The best stage of maturity is determined by head size and condition. The desirable characteristics are white coloured compact head with smooth surface, tender texture, absence of looseness and not too thick flower stalk. Over mature head becomes too long, flower stalks elongate, resulting in loose, leafy and fuzzy conditions.

Cucumber

Slicing cucumbers must be medium sized, dark green, immature with small seeds. Most of the varieties are harvested from 16 to 23 cm long. Pickling cucumbers are harvested earlier than slicing cucumbers. The usual length of pickling cucumber is 5 cm. A greenish colour is desirable.

Garlic and Onions

In green onions harvesting takes place 45 to 90 days from field settings and 90 to 150 days for bulb variety.

Bulbs mature when the neck tissues begin to soften and the tops are about to abscise and decolorize. Development of red pigment and characteristic pungency of the variety are also important harvest indices of onion. Regarding garlic, cloves are ready for harvest at 100 to 140 days after field setting.

Radish

The crop matures from 3 to 4 weeks from field setting in quick grading varieties and 8 to 14 weeks for Chinese varieties. At this stage they are mild, tender and crisp. They are harvested before they become pithy and fibrous.

References

1. Peal Engineering, 2003. Vegetable handling systems http:// www. pealeng.com /graders/ potato/ index.phtml on 3.12.2003.
2. Ramanna, K.V.R 1993. Harvest indices for fruits and vegetables. Paper presented in the Advanced Technology Training Programme, 27 September, 1993, CFTRI.
3. Rodriguez, R. 1993. Post-harvest management of fruits and vegetables. Paper presented in the Advanced Technology Training Programme, 27 September, 1993, CFTRI.

4. Shrivastav, S. 2004. Post harvest handling of horticultural produce and technology adaptation. *Processed fd. ind. 7:* 22-36.

5. Wills, R.B.H., Lee, T.H., Graham, D., McGlasson, W.B., Hall, *E.G.* 1981. Post harvest: An introduction to the physiology and handling of fruits and vegetables. AVI Publishing Co. Inc. Westport pp.142-145.

Chapter 3

Washing

For getting a clean produce, fruits and vegetables have to be washed after harvesting. It improves the appearance of the produce. In order to remove soil, insects, sooty moulds and also the pesticides and fungicide residues, washing the produce is an inevitable process. Washing in cold water can also reduce the field heat hence the storage life is prolonged. For example, there is report that washing of bananas delay the onset of the climateric peak.

Hot water successfully eradicates incipient infections in several fruits. For example, dip in hot water (50°C) for 1–2 minutes inactivates the infections of *Phytophthora* sp. in mature green tomatoes, orange and lemons, *Diplodia* in citrus, *Colletotrichum* in papaya and mango and crown rot of banana. Hot water is the best heat transfer medium because of availability, heat capacity and lack of residue in the fruit.

Washing Equipments

The field produce has to be thoroughly washed and graded before packaging into the containers.

Hand Washing

When ever large quantities of fruits and vegetables has to be handled, it may not be possible to resort to hand washing. In such occasions we need suitable mechanical devices to carry out the operation mechanically. Here are some of the commercially employed equipments suitable for each type of materials.

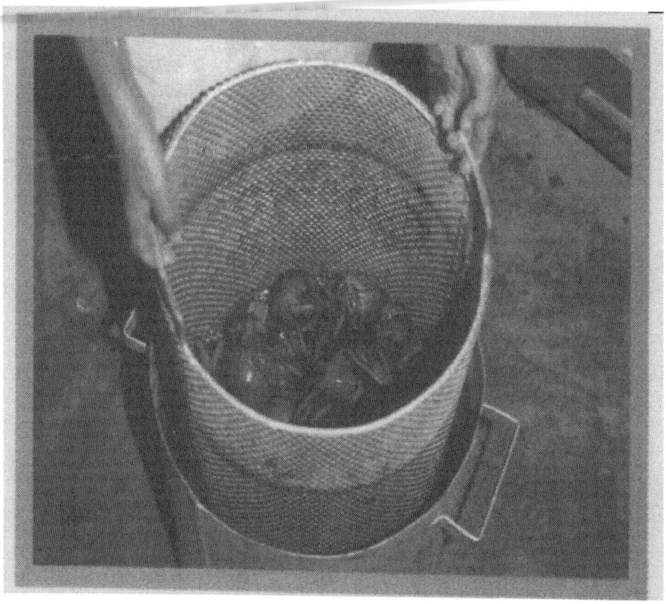

Figure 3: Fruits and Vegetables for Washing is Taken in the Inner Vessel having Full of Perforations and Repeatedly Immersed in the Outer Vessel Containing Water with 30 ppm Chlorine. Quantity taken in the inner vessel is that sufficient enough to be handled by a person.

Soak tank Washing

Generally the produce has to be soaked in water for 3 minutes to loosen the dust, microbes and pesticides etc. This is generally done by soaking the produce in soak tank where water is circulated and the produce is agitated. The filtering of water is also done during circulation to remove the dust particles. Sodium hypochlorite solution may be added into the soak water to control possible buildup of thermophilic microbes. A residual chlorine content of 6 to 8 parts per million may be maintained at all times. The soak water is generally chilled to remove the field heat.

The produce is generally dumped into the soak tank and removed by the belt conveyors having perforations to take out the produce without water for next stage of cleaning. The detergent solution can also be added in the soaking water if required. In some of the soak tanks, rotary brushes are fitted to clean the produce and these are then known as brush washers. When the paddle wheels are fitted for agitation and cleaning of the produce these are then known as immersion washers.

If the produce is for processing, the second stage washing is done in rotary barrel type washers and then washed by spray washers. If the produce is for packaging, the produce is being cleaned by spray washers only after it is brought from the soak tank.

Rotary Barrel Washing

The rotary barrel washer is having a capacity from 1 tonne per hour. It consists of a cylindrical barrel with holes on periphery, the barrel is generally lined with rubber material for cushioning. The agitators are also fitted for

Figure 4: Soak Tank

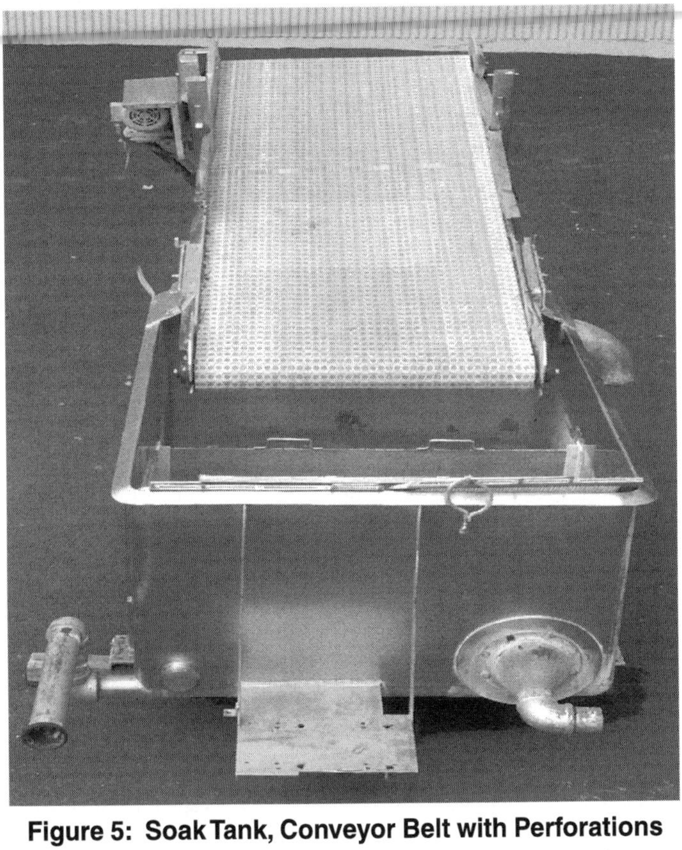

Figure 5: Soak Tank, Conveyor Belt with Perforations

Figure 6: Rotary Barrel Washer

Figure 7: Rotary Barrel with Holes

proper agitation of the produce. The barrel is partially submerged in water and water is circulated and sprayed over the rotary barrel to assist in cleaning. The produce from the soak tank is fed to the rotary barrel washer where it rotates and tumbles the produce in water. The external

Figure 8: Rotary Barrel with Agitator

impurities get removed from the produce by rubbing of the produce with each other and the lining of the barrel. The water is circulated and filtered. The water is changed when required. In some of the models, the cylindrical rotary brush is provided in the center which is also rotated and help in cleaning.

Spray Washing

The produce from soak tank or barrel washer is conveyed to spray washer. It consists of roller conveyor and spraying nozzles. The individual fruit is made to rotate and water is sprayed under pressure over this by nozzles. The washed fruits are then sent to sorting table. It generally consists of roller conveyors where the produce rotate during

Figure 9: Spray Washer

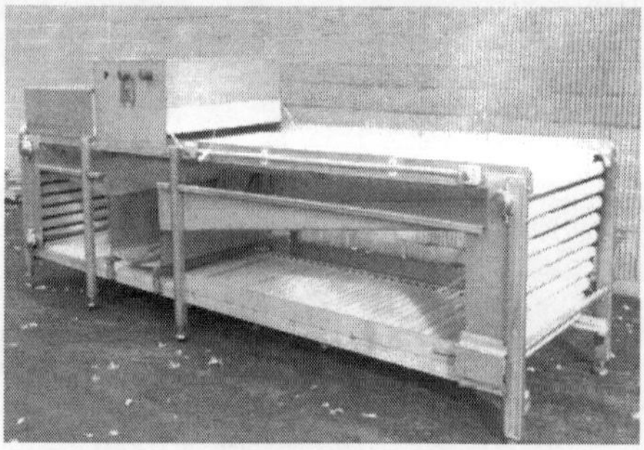

Figure 10: Spray Washer cum Inspection Table

conveying. The culled fruits and unclean fruits if any, are picked manually and discarded.

Brush Washing

When the soak tanks are fitted with rotary brushes to clean the produce they are known as brush washer. The washing procedure here is same as that in soak tank. The rotary brushes provided will make an intimate thorough cleaning of the surface of the materials and is an ideal washer for commodities of hard nature.

Hot Water Washing

Produce may be immersed in hot water before storage or marketing to control diseases. Fruits after harvest dipped

Figure 11: Brush Washer

in hot water (45-55°C) for about few minutes for uniform and rapid ripening and for control of decay. A common disease of fruits, which can be successfully controlled in this way, is anthracnose caused by *Collectotrichum* sp.

Many fruits and vegetables tolerate hot water temperature of 50-60°C for up to 10 minutes but shorter exposure at these temperature can control many post harvest plant pathogens. Low concentration of fungicides can be applied along with hot water treatment, thus allowing more effective control with a reduction in chemicals. Heated solutions (45°C) of So_2, ethanol and sodium carbonate can be used to control green mould *(Penicillium digitatum)* on citrus fruits.

Similarly hot water spray machines are also developed for control of decay.This is a technique designed to be a part of the sorting line where by the commodity is moved by means of brush rollers through a pressurized spray of hot water. The exposure time of the commodities with hot water will depend by varying the speed of the brushes and the number of nozzles spraying the water. Commodity can be exposed to high temperatures for 10 to 60 seconds. The water is recycled. This type of machine is currently used in Israel both to clean and to reduce pathogen population on a number of fruits and vegetables.

Hot water dips have also been tested for their efficiency in disinfesting insects. The time of immersion can be one hour or more at a temperature below 50°C, as against many antifungal treatments that require only minutes at temperature above 50°C. Fruits like apricot, nectarine and peach can be cleared of thrips by 2 or 3 minutes in hot water at 48 to 50°C.

References

1. Dobmac 2003. Grading, packaging and washing equipments. http www. Dobmac.com.au on 12.9.2003.
2. Fallik, E., Aharoni, Y., Yekutieli, O., Wiseblum, A., Regev, R. Beres, H. and Bar Lev, E. 1996. A method for simultaneous cleaning and disinfecting agricultural produce. *Israel Patent Application No.* 116965.
3. Kader, A. 1980. Keys to successful handling of fruits and vegetables. *California conference on direct marketing models,* California. P. 1–5.
4. Wills R.B.H., Lee, T.H., Graham, D., McGlasson, W.B., Hall, *E.G.* 1981. Post harvest: An introduction to the physiology and handling of fruit and vegetables. AVI Publishing Co. Inc. Westport pp. 142-145.

Chapter 4

Sorting and Grading

Importance of Sorting and Grading

Sorting and grading in fruits and vegetables are developed to identify the degrees of quality in the commodities, which aid in establishing their usability and values. These are important tools in the marketing of fresh fruits and vegetables. Sophisticated marketing systems require precise grading standards for each kind of product. Primitive markets may not use written grade standards but the products are sorted and sized to some extent. The main purpose of sorting or grading is to segregate produce into grades A, B, C and culls. Over ripe fruits or over mature vegetables, unpalatable for human consumption may find use in animal feeding or in processing of some products. The primary properties of fruits and vegetables that are used in typical sorting operations are shape, size, colour, ripening degree, mechanical injuries etc.

In the developed countries almost all agricultural commodities are marketed on the basis of official standards established under federal or state laws. In the present day marketing system pricing is tied to produce quality. Adopting uniform commercial code is of paramount importance in the distribution process. It indicates that farmers, agents, traders and retailers use the same term or logo while dealing with a particular commodity, its size, colour and other quality factors.

Sorting and grading should be viewed to include all operations that segregate a material with a mixture of attributes in to distinctive groups (grades or size) whereby the concentration of the material with particular attributes is much larger than in the raw material.

In many textbooks, sorting and grading are generally taken to mean the same thing and are used interchangeably. But a slight distinction has been made between the two words, because in India sorting is generally resorted to and grading is not so common. Here, in the sorting of fruit and vegetable sensory quality attributes are taken into consideration, such as presence and extent of discolouration, blemish and shape; while in grading physical attributes which can be measured such as size and weight are taken into account. Often sorting precedes grading operations.

The simplest sorting operation comprises a raw material which contains only two different fractions. They are usable commodity and thrash like leaves, vines etc, normal fruit and undersized fruit or good fruit and decayed fruit. The differences in physical properties (attributes) of the good fruit and culls are used as means to distinguish between them in the segregation process. A similar situation exists when a pre-sizer in an apple packing house separates undersize apples, which are unsuitable for fresh consumption from the larger apple.

Produce is graded by human eyes and hands, while moving along conveyor belts or rollers. "Electric eyes" are some times used to sort out the produce by colour. In small scale packing operations, one or a few grading tables may be enough. Typical grading facilities in large packing houses include dumpers and conveyors. Dumping, conveying and grading can cause mechanical injury to some products. Equipments should have a smooth, soft surface. Dumping and grading operations should be gentle, to minimize injuries.

Advantages of Sorting and Grading

1. It provides common language to producer, buyer and consumer
2. Reduce dispute of quality between seller and buyer.
3. Standardized grades form basis for price fixation and advertisement.
4. Improve marketing efficiency by selling a produce without a personal selection.
5. Assist producers and other intermediaries in preparing fresh horticultural commodities for market with appropriate labelling.
6. Provide a basis for securing incentive price for better quality.
7. Serve as a realistic basis for market intelligence and reporting.

8. Prices and supplies quoted in different markets could only be meaningful if they were based on products of comparable quality/grade.

Additional Benefits

1. Bargaining power is enhanced to get better premium for better grades.
2. Grading the produce into certain size and shape facilitates packing and transportation easy.
3. To eliminate re-sorting or re-packing at wholesale or transit markets.
4. Grading can develop specific market for a specific type/variety or produce

Many products are sized according to their weight. Automated weight sizers of various capacities are used in packing house. Round or nearly round fruits are often sized according to their diameter, using automated chain or roller sizers or hand carried ring sizers. An inefficient sizing operation can also cause significant injuries.

Some of the factors responsible for differences in quality of produce are genetic, while some others are environmental or agronomical. Examples for defects caused by environment are cracking and splitting in tomatoes, presence of green shoulder at ripe stage in some cultivars of tomato. Blemishes caused by unfavourable environmental conditions also lower the quality. Green color near the crown in carrots, green portions in potatoes due to exposure to sunlight and sun scald in tomatoes are some examples of this type, commonly seen in the handling of fruits and vegetables. Natural enemies such as insects, disease, hail and wind also cause blemishes, scars and other defects. Many crops are naturally variable in size and other factors associated with maturity, such as colour and firmness. Mechanical injuries received during mechanical harvesting or during transportation are also responsible for variation in quality in a lot.

Standardization in terms of grade consists of several quality factors such as colour, shape, size, weight, maturity, freshness, ripening, blemishes etc. A well designed programme of grading and standardization brings about an overall improvement not only in the marketing systems, but also results in quality consciousness among producers, traders and consumers in the long run. Further, systematic grading coupled with scientific packaging and storage, reduces the post harvest losses and marketing costs substantially and enables the producers to fetch a competitive price. Therefore, it is obvious that, sorting and grading is an essential pre-requisite for efficient marketing system.

Grade Standards for Fruits and Vegetables

The Directorate of Marketing and Inspection (DMI) of the Ministry of Rural Development, Govt. of India, is the pioneer organization in the country in formulation of grade standards (Agmark grades) for agricultural and allied commodities under the provisions of the agricultural produce grading and marking Act 1937. It has also been entrusted with implementing the same in the field. In respect of fruits and vegetables the DMI has framed grade standards both at producer's level at the time of sale and at the consumers level before sale of the commodity.

As regards producer's level grading, grade standards are derived and designed in such a way to suit local conditions. In this system, normally, the commercial grading is done either as per AGMARK standards or the liberal standards laid down by the State Governments on a voluntary basis without involving the time consuming quality control practices such as, packaging, labelling, marking and sealing which are invariably followed for grading under AGMARK. The basic objective of this system of grading is to quicken the process of sale on the basis of grades and to reward the producer with a price commensurate with quality of produce. The DMI has commenced its grading activities in different phases. Voluntary grading of centralized commodities started in the year 1936-39, decentralized commodities in 1948-49, compulsory grading in 1942-43 and voluntary grading at producers level in 1963-64. The important disadvantage is that sizing and grading manually requires labour.

Standardized Grades of Some Fruits and Vegetables

Citrus grade (NRCC, Nagpur)

Big – 8.5 –7.5 cm

Medium –7.49-6.5 cm

Small – 6.49 – 5.5 cm

Figure 12

Mango – Desheri (DMI)

Figure 13: Excellent

Poor– <130g

Average –131-170 g

Good – 171–200 g

Excellent– >200 g

Cabbage and Cauliflower (DMI 1997)

Figure 14: Large

Small– <1 kg

Medium–1 –2 kg

Large–>2 kg

Pineapple (DMI)

Figure 15: Class A

A class–1500-1800 g

B class–1100-1500 g

C class–900-1100 g

D class–700-900 g

Potato (C.P.R.I. Shimla)

Small–< 25 g

Medium–25-50 g

Large–50-75 g

Extra large–> 75 g

Figure 16: Medium Size

Progress of Grading of Fruits and Vegetables in India

Quality and value of fruits and vegetables are graded on compulsory basis for export, voluntary basis for internal marketing at producer's level. It may be observed that, only table potatoes and onions were compulsorily graded under AGMARK for export in the country prior to liberalization of country's economic policy.

In the earlier period the produce from the same basket is sorted into separate lots and priced differently depending on the quality, grade and status of maturity. This concept is lacking in most of the growers who do not care to grade and pack their produce in accordance with quality and size. Therefore the farmer gets paid as per the lowest grade of the produce prevalent in that lot. Besides, this also necessitates the buyer inspecting each pack for quality before purchasing. So special efforts were made to promote the concept of delivering high grade fruits to the consumer.

In formulating standards, terms defined in a general descriptive forms are like colour, shape, freshness and firmness. Followed with the terms for serious defects like decay, insect damage, freezing injury etc. Less serious defects are ordinarily covered by expression such as free from damage or free from serious damage depending on the grade designation used. Since it is not humanly possible to remove all defective specimens as a product moves over the sorting line, tolerances for grade defects are written in to the grade standards.

In India the need for grading agricultural and horticultural produce was realized decades ago when the agricultural produce (Grading and Marking) Act was enacted in the year 1937. This Act conferred statutory power to the Central Government to frame grade standards for various agricultural and allied commodities and notify the same; under this Act the first commodity for which grading and marking rules were framed was grapes. Subsequently, the Directorate of Marketing and Inspection framed grade standards, popularly known as "AGMARK STANDARDS", for a number of fruits, both for export purposes as well as for domestic marketing and notified grading and marketing rules in respect of mangoes, apples, oranges, sweet limes, sour limes, pears, plums, grape fruit, lemon, potato etc. Grading under this rule is voluntary. Thus produce graded under "AGMARK" have to follow the statutory procedures and obtain license known as "Certificate of Authorization".

The grade specification provided under the Agricultural produce (Grading and Marking) Act 1937 prescribed parameters for special and general quality characteristics for different fruits and vegetables, while the special characteristics prescribe certain variables like minimum length, weight, diameter etc. Of the fruits which are measurable, the general characteristics include attributes such as uniform shape, colour, size, stage of maturity, freedom from defects and disease and good keeping quality. Fruits like mangoes, oranges, pineapple, lime and lemons, chickoo, grapes and apples are graded under AGMARK for consumption within the country. This sort of grading is, however, mainly consumer oriented and did not make any significant impact on the marketing system at the primary level.

In India like mangoes and potatoes, the grading system adopted for the Nagpur Mandarin fruits are on their diameter base. Following are the grade standards developed for the following produces:

Table 7: Potatoes

Grade	*Tuber Size*	
	(Minimum Diameter) Oval or Long Varieties (mm)	*Round Varieties (mm)*
Extra special	41 to 89	45
Special	29 to 83	32

Potatoes should be reasonably clean, healthy, free from serious defects and suitable for human consumption.

Table 8: Mangoes (Variety Alphonso)

Grade	*Weight of Each Fruit*		*Definition of Quality*
	Min(g)	*Max(g)*	
I	280	338	1. The fruits shall be firm and entirely free from damage, blemish or malformation
II	222	280	
III	163	222	2. Each fruits shall be olive green in colour without trace of yellow at the time of packing

Table 9: Mandarins

Size Grade of Mandarin	Size Variation (diameter)
Big	7.50–8.50cm
Medium	6.50–7.49 cm
Small	5.50–6.49 cm

To allow for accidental errors in grading, a tolerance of 5 per cent may be allowed in respect of fruits which are above or below the limits of weight specified.

Central potato research institute, Shimla has developed hand operated potato grader, power operated potato grader and continuous potato grader with conveyor for the purpose of sorting of potatoes into different grades.

In spite of these developments, mechanical grading of fruits and vegetables in India is still in the infant stage, and the grading equipment has not been put into large scale commercial use in the field.

Measurement of Quality in Grading

In the beginning most of the product characteristics which affect grade are determined subjectively by the human sense. He/she estimates colour, shape, firmness and freshness largely by eye and touch, based on the training and experience he/she has received. However these factors to great extent were usually quite accurate.

Instruments are also developed for mechanical or electronic sorting of products on the basis of external or internal colour, internal defects or even physiological maturity.

Instrumental measurement of texture has application in the processing industry; for example the shear pressure for deciding the texture and maturity of peas, corn, asparagus etc. Magness Taylor pressure tester indicates harvest maturity and ripening level of fruits. The 'Instron' a universal testing instrument allows measurement of shear resistance (example to measure fibre strength in celery)

Specific gravity is widely used as a measurement of quality in potatoes. Buyers of potatoes for manufacturing chips buy largely on the basis of specific gravity or total solids. Specific gravity also provides a means of separating potatoes on quality basis. The usual way is by floatation in saline solutions of different densities. Measurement of total sugars is another indication of quality in fruits, which can be easily measured by a hand refractometer.

Colour Grading

Until rather recently, all colour standards for grade or quality were based on visual estimation. Photoelectric colour sorting machines are now available. They will separate fruits into several classes from green to yellow (lemon), green to orange (oranges) or to several shades or amounts of ground colour (apples). Grading is accomplished by measurement of the light reflected from the product. Colour separation mechanism being activated by the quantity and quality of light received. The unit is reported to be used commercially for colour sorting lemons for fresh fruit market and red tart cherries for processing.

One of the first efforts towards objective measurement of colour involves the use of light reflectance. Hunter colour difference meter (HDM) is now used widely by tomato processors for establishing the value of individual lots of tomatoes for processing in to juice, ketchup and paste.

It is also possible to separate apple into quality grades using fruit colour. Golden delicious apples at harvest can be sorted into different groups according to the extent of chlorophyll content as measured with a difference meter (I.Q sorter).

Sorting for Internal Defects

With a similar technique to that used for measuring chlorophyll content in the interior of fruits, light transmittance can be used for the non-destructive detection of some internal defects in fruits such as presence of water core, except in very severe cases, water core can be detected only by cutting fruits to determine its presence and severity. Now water core can be detected non-destructively by several methods like comparative specific gravity technique or by using light transmittance technique. The best measurement of severity of water core is by light transmittance at OD 760 nm to 810 nm.

Grading for Size

Sizing of produce, *i.e.*, separating produce into size groups is defined as grading by size. Fruits and vegetables grow in nature in different size range. Size range differs widely according to many factors such as biochemical, climatic and agronomic practices. Individual samples may differ considerably from a normal binomial distribution pattern. Since some orchards produce predominantly large while others may grow predominantly smaller fruits. Normally, only the commodities falling in the center section of the distribution curve is marketed as fresh produce, while produce outside the marketable size range is considered as culls for animal feed or converting to other products. Oranges are typically marketed in size range of 50 to 90 mm while a pre-sizer usually eliminates those below 45mm. Extreme averages are generally not marketed.

Manual Sizing or Size Grading

Before the advent of sizing machines packers used to 'size' the produce manually by picking the appropriate size from a pile of fruits.

The whole packing process consisted of grading and sizing simultaneously by 'seeing' and 'feeling' the produce selected. This type of visual sizing is quite good and as a matter, it is better than modern sizing machines. As workers primarily use sight and touch (firmness) for hand grading, these senses can be used most efficiently in grading crops large enough to see well and to handle easily by hand. Sizing of many vegetables is based on human judgment and is a manual operation. It may be done by graders who select the grades on conveyers to be carried to bins for the packers. This method is used for melons, carrots and sweet potatoes.

Equipment for Hand Grading

In small simple operations, a crop may be spread out on a flat surface and sorted satisfactorily by hand. However the side portion of the sorting equipment can help workers to reach his/her full potential for sorting. The main piece of equipment for manual grading is the sorting table. The commodity, which is to be sorted generally, moves over a belt or roller conveyor. The roller conveyor turns the product as it moves forward and sorters are able to see all sides of each item. The commodity is graded in a continuous flow.

There are two types of operations in manual grading. (1) Reduction grading and (2) Full grading. In reduction grading the sorter picks out one or more grades which comprise the minority of the fruit flow (unsuitable, non-confirming to the grade), while the predominant grade is allowed to flow on to the next processing section. If the percentage of grade A is less than grade B then grade A should be picked out while grade B is allowed to flow on. This reduces the total effort per tonne of graded produce.

Full grading consists of picking up all the produce off the sorting table and placing it on the appropriate conveyor, thereby forcing the sorter to inspect every single fruit or vegetable. Full grading is much more efficient than reduction grading but involves much labour and therefore expensive, hence seldom used in large scale packing house.

Feeding rate in the conveyors must be controlled to match the workers ability to grade and to achieve the desired degree of precision; even the flow and distribution of produce on the sorting belt improves worker efficiency.

Dividing the sorting table in to lanes and assigning a lane to each sorter improves sorter efficiency and responsibility.

Good visibility of a product is important for efficient hand grading; therefore proper lighting is an important factor.

Machine Sizing

Mechanical sizers are commercially available; sizing machines may involve only elimination of those units, too small or too large for mechanical acceptability.

Equipment for Machine Sizing

Sizing by Weight

Weight sizers are used principally for long cultivars of potatoes, cucumbers and sweet potatoes. Weight sizing particularly with product of non-uniform shape, relates length to diameter more accurately than sizing based on

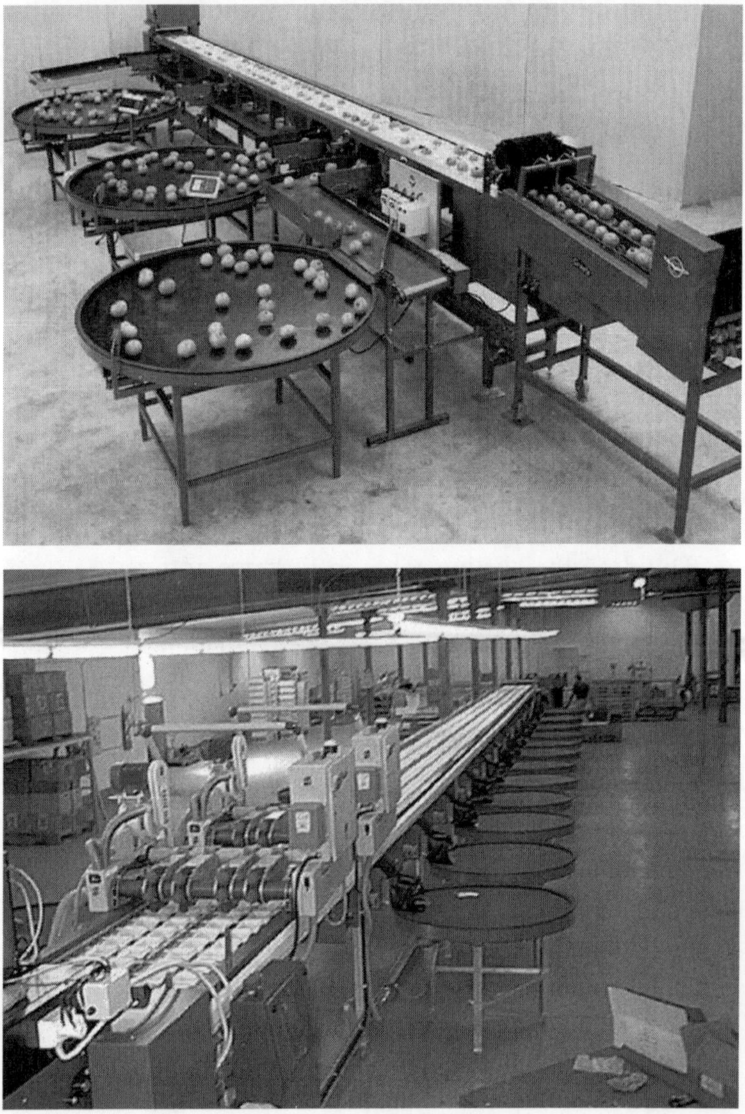

Figure 17: Weight Graders

Figure 18: IIHR Electronic Weight Grader
1: Grading unit; 2: Singulation unit; 3: Control unit; 4: Weighing unit

Figure 19: Singulation Unit

length or smallest diameter. A widely used weight sizer has a turn and timing table which aligns tubers or other product before delivering them into the weighing cups, which are gently vibrated to ensure centering for accurate weighing. Each potato is weighed as it moves along the sizer assembly. As it reaches the weighing area the cup is tipped and the tubers falls on to a side delivery belt for delivering to a packing bin.

Mechanical Weight Graders

In this grader, there is an endless chain of cups.The grader may have one to ten channels to convey the fruits. Different kinds of cups are used to hold and grade fruits.The mechanical weight sensing mechanism is fitted below the chain in the form of cups which are carrying the fruits. The fruits are fed into thc cups by the conveyors and singulation unit. The fruits move in the cups all along the chain. The fruits with larger weight get graded near the feeding end by tilting of the cups when the weight of the fruit is more than the tipping / holding capacity as per the adjusted capacity of the sensing unit. As the remaining fruits move along the length of the conveyor, lesser and lesser weight fruits get graded and collected in different grades as per the adjustment of the sensing unit. The fruits of different grades are then pass on to the sorting tables.These are then packed in Corrugated Fibre Board (CFB) boxes manually by labourers and any cull or damaged fruits are sorted out. The weight of the filled CFB box is maintained as per the requirement and the box is then closed. In packing apple and other high value produce, the cushioning material like pulp tray or 'styrofoam' trays are also used and the required number of fruits per CFB box is filled. In some of the packing houses the fruits of the different colours are sorted out manually and packed in different boxes for the same weight grade.

Sizing by Diameter

Diameter sizers are commonly fabric or steel mesh belts with holes for products of certain sizes to drop through; they may have only one size hole, or they may consist of sections with smallest holes in the first section and increasingly larger holes in succeeding sections. A cross belt conveyor under each sizing section carries the product to the packaging bins. This type is commonly used for sizing mature green or pink tomatoes.

Many diameter sizers use rollers, either longitudinal or crosswise, which either decrease in diameter (longitudinal rollers) with increasing distance from the intakes end or retake at increasing distance from each other as the product progresses. Most of these rollers operate as conveyers as well as sizers with the smaller items dropping first and the larger going through as the distance between rollers increases. The rollers may be in the form of brushes or of solid wood or steel. Some of these diameter sizers have cross conveyor belts while others drop the product in to chutes which feed directly in to the packaging bins.

Size Graders

Oscillating screen graders were used to screen of under size produce, however, these have been modified as follows.

Screen Graders

The screen graders consist of more number of screens moving at constant speed. The smallest size of the screen is fitted at the feeding end and biggest size near the outlet

Figure 20(a): Screen Grader

Figure 20(b): Screen Magnified

end. The produce is continuously fed by elevator conveyor to the grader feeding end and the grades are collected in the respective outlets. It is suitable for guava, onion and other fruits and vegetables.

Fixed Opening Roller Type Graders

These graders consist of the endless belts, feed hopper and a frame. The two endless fixed roller belts having regular opening of size 40 cm x 5 cm and 40 x 2.5 cm move one over the other with the same speed and the same direction with gentle shaking. The shaking helps in changing the orientation of the products thus increasing the grading efficiency. The ungraded material from the feed hopper being fed to one end of the upper belt, the product of the size larger than 50 mm is carried over by the belts to the other end of the belt where it is discharged into the bags attached with the holder or for packing into other packages. The product smaller than 50 mm size falls on the lower belt which has 25 mm openings. This belts retain the product of the size between 25 mm and 50 mm and allows the product of less than 25 mm to fall through the gap

Figure 21: IARI Potato Grader

between the PVC rollers on lower belt. This graded material is collected separately. The medium size product is carried away by the lower belt, discharged and collected at the end of the lower belt. This machine separates the product into three grades of desired size. The belts of different spacing of PVC rollers can be used for sizing the product. This grader is suitable for grading potatoes and it needs ¼ hp electric motor to operate. The optimum belt speed is 1.6 m/s and the capacity is around 1.8 tonne per hour.

Rotary Barrel Grader

In this machine, a cylindrical barrel is provided with holes or slots for grading round or cylindrical fruits fed from one end, the barrel is rotated around the central axis, the fruits smaller than the opening passes through this. The produce is thus graded into grades. (*i*) Small size and (*ii*) Bigger size. The barrel grader is used to grade sapota. The machine needs around 1 hp motor and the grading capacity is 2 tonne per hour. Sometimes, the fruits clog the openings and are to be dislodged. This again falls in the

Figure 22: Rotary Barrel/Drum Type Grader

barrel. The damage to the fruit is more than the other types of grader.

Divergent Aperture Roller Type Grader

This grader is used for grading fruits into three sizes. Here the fruits are fed into rollers which are having expanding opening and rotating in opposite directions. The fruit moves in the divergent opening due to rotation of the

Figure 23: Divergent Opening (Roller) Two Stage

Figure 24: Divergent Opening (Roller) Single Stage

rollers. Fruits will pass through the opening when the size of opening is equal to size of the fruit. This kind of grader is most suited for round shape fruits like apple, orange and conical vegetables like carrots.

Expanding Opening Double Belt Grader

This consists of two endless belt sets moving in same direction. The distance between the belts is increasing along the length and the belts inclined to the vertical by 35° to 45° to hold the fruits.The ungraded produce is fed from the end having smallest opening. Fruits get singled and get held in the space between the belts and move along the direction of motion. Small size fruits get graded by passing through the small opening near the feeding end. The medium size fruits get graded by passing through the medium opening. The big sized fruits get graded by passing through the largest opening near the far end belt. The graded fruits can be collected at the respective outlets. The grades can be adjusted as per the number of openings. This grader is suitable for round fruits and can be adjusted for different size by adjusting the opening between the belts.

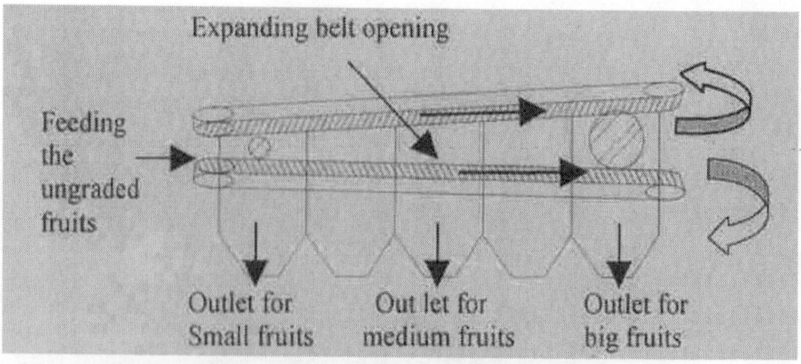

Figure 25: Expanding Opening Double Belt Grader (PAU, Ludhiana)

Expanding Opening Single Belt Grader

One endless belt is moving along a channel created for the movement of the fruits by fitting planks on the side. One side plank is having the divergent opening, small opening near the feeding end and large size opening near the far end. The fruits are fed near the small opening and is allowed to move along the belt. Small size fruits pass through the opening near the feeding end and medium size fruits get graded through the medium size opening near the center of the belt and large size fruits get graded at large size opening near the far end of the belt. This grader is most suitable for round fruits like oranges.

Figure 26: Expanding Opening Single Belt Grader (CIPHET, Ludhiana)

Expanding Spool/Roller Grader

This grader consists of frame, an elevator, feed conveyor, an intermediate receiving conveyor, a sizing conveyor with rubber spools / rollers and two identical driving rollers with helical groves of gradually increasing pitch.The helical groove rollers with increasing pitch are fitted along the length of the grader. The rods with the spools / rollers are

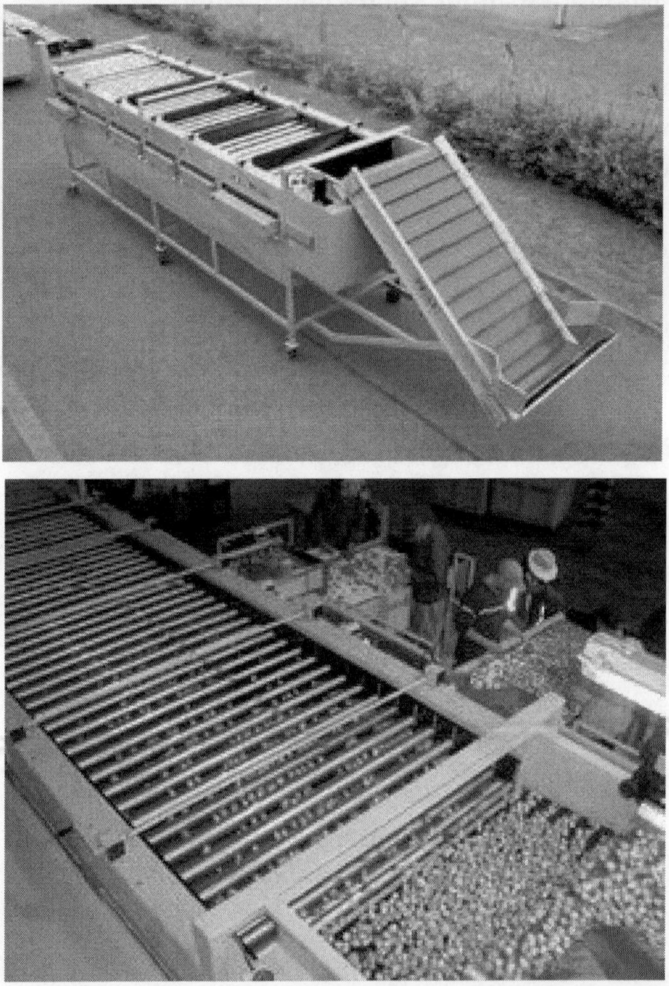

Figure 27: Expanding Opening Roller Grader

carried forward as the driving helices rollers rotate. The gap between the individual spools/rollers goes on increasing gradually according to the pitch of the helices of the driving rollers. This leads to sizing of fruits as these are carried forward over the sizing conveyors. The produce can be graded into different grades by proving the necessary collecting outlets / conveyors below the spools / rollers. This type of graders are extensively used commercially for grading apple, orange, onion, potato etc.

Expanding Opening Slot Type Grader

This grader consists of a variable flat spacing chain conveyor for grading of fruits. Along the length of the conveyor the gap between flats (made of stainless steel) increase to drop through the smaller fruits first and bigger fruits later. One end of the stainless flat was hinged to a rod connecting two tracks of the chain. The lower end of the flat is attached with resting wheel and these resting wheels move on a track. The height of track can be adjusted by using appropriate mechanism to make the grader suitable for different kinds of fruits. This type of grader is being extensively used for grading sapota.

Figure 28: Onion Grading

Computerized Graders for Size, Weight and Colour

The fruits can be graded as per size, weight and colour or a combination of these. The size of the fruit is sensed by the camera and the colour of the fruit is sensed by the computer. For determining the different colour grades, about 10 fruits of required grade are fed on the roller conveyor chain underneath the camera of the computer for calibration. The weight of fruits is also sensed near the feeding end. The different required grades for size, weight and colour are fed to the computer. The fruits move on the chain with cups as explained in mechanical weight grader. These get discharged into the different outlets for different grades as per the data fed to the computer. For colour

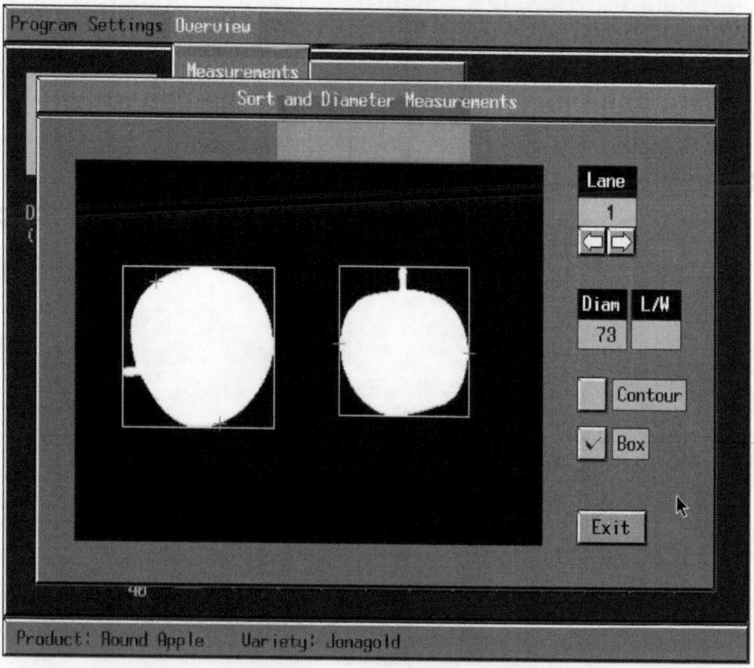

Figure 29: Size Sensing

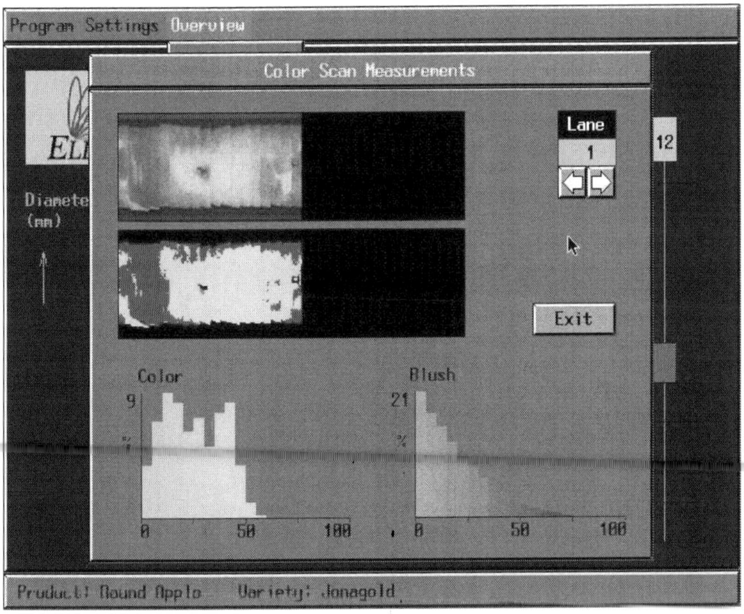

Figure 30: Colour Sensing through Cameras

grading roller conveyor are used. The computer can also discharge equal number of fruits for same grades in different opening if required. The different grades get collected into different outlet. These are then filled manually into corrugated fibre board (CFB) boxes. The computer can also give the signal at the required outlet. The colour grader can also discard the fruits with unwanted colours and spot diseases and mechanical damage.

References

1. Causter France, 2003. 'Grading and packaging of fresh fruits, vegetables. http:// www. caistier. com/ Andleterre/grading/summary.htm 27.11.2003.

2. Ellips, 2003. Fruit grading system, http://www.ellips.nl/oem/fruitgrading.h 27.11.2003

3. George and Courtier, Pvt. Ltd., Brishana, Australia. 1990. Pamphlets on harvesting, handling, sorting, grading equipments and post harvest treatments.

4. Haith Tickhill Group Companies, 2003. Vegetable grading machi http://www.haith.co.uk/grading.html on 27.11.2003.

5. Rodriguez, R. 1993. Sorting and grading of horticultural produce. Paper presented in the Advanced Technology Training Programme, 27 September, 1993, CFTRI.

Chapter 5
Pre-treatments

After sorting and grading thc most important opcration next is the pre-treatments prior to packaging, all the pre-treatments are not required in all the cases, it will vary depending upon the commodity and purpose, however pre-cooling is an important pre-treatment in the protocol of any post harvest management of fruits and vegetables. Some of the pre-treatments discussed here are curing, irradiation, skin coatings, pre-cooling, heat treatments etc.

Curing

Many root crops have a cork layer over the surface which is called periderm. This serves as a protection against microbial infections and excessive water loss. This layer can be broken or damaged during harvesting and handling operations, so curing is essentially a wound healing operation to replace the damaged periderm mainly done for tuber crops like potato, sweet potato, yams etc. Curing is done for onion and garlic for proper colour development, reduction of moisture content etc.

Mostly done in the case of potatoes and onions. Early onion crops produced in southern and south western states are seldom completely mature when harvested. Leaves are often green and wet and susceptible to infection. The usual practice is to leave the onions in fallows a few days before topping and then lining or stacking the bagged or crated onions in the field for a few days before transferring to the packing house. This is a satisfactory procedure if the weather is dry and the relative humidity is low during field curing.

Blowing heated air at 43°C to 46°C vertically through a grill on which onions in mesh bags have been placed is one of the successful artificial methods of curing adopted in onions. Such treatments will be continued for a period of 8 to 10 hrs to get a satisfactory curing for either immediate shipment to market or to storage for later sale.

Curing of root, bulb and tuber vegetables offers effective means for reducing post harvest decay. The operation is conducted immediately after harvest and generally done at the farm level itself. The process involves suberisation of outer tissue followed by the development of wound periderm which acts as an effective barrier against infection and water loss.

In the case of yams, the bruised and cut injuries of yams can be cured by exposing the tubers to the sun for a short time or by prolonged drying in warm and well ventilated shed.

Irradiation

Irradiation processing of food involves the controlled application of energy from ionizing radiation such as gamma rays, electrons and X-rays for food preservation. Gamma rays and X-rays are short wavelength radiation of the electromagnetic spectrum which includes radio waves, microwaves, infrared, visible and ultraviolet rays. Gamma rays are entitled by radioisotopes such as Cobalt 60 and Caesium-137, while electrons and X-rays are generated by machines using electricity

Applications

1. Inhibition of sprouting
2. Longer retention of quality
3. Insect control
4. Delay in ripening

Ionizing radiation can be applied to fresh fruits and vegetables to control microorganisms and inhibit or prevent cell reproduction and some chemical changes. High energy short wave gamma rays are used. It inhibits sprouting of onion and potato, also helps to control pest and disease which are very difficult to be controlled by chemicals. Delay in ripening, longer retention of high quality etc are some of the other effects of irradiation.

Irradiation processing of food involves exposure of food to short wave energy to achieve a specific purpose such as extension of shelf life, insect disinfestations and elimination of food borne pathogens and parasites. In comparison with heat or chemical treatments, irradiation is considered more effective and appropriate technology to destroy food borne pathogens.

Table 10: Various Applications of Irradiation with Respective Doses

Name of Food Material	*Purpose*	*Dose (KGy)*	
		Minimum	*Maximum*
Onion	Sprout inhibition	0.03	0.09
Potato	,,	0.06	0.15
Ginger	,,	0.03	0.15
Garlic	,,	0.03	0.15
Shallots (small onion)	,,	0.03	0.15
Mango	Disinfestations	0.25	0.75
Fruits	Delaying fruit ripening	0.25	0.75
Rice	Disinfestations	0.25	1.00
Raisins, fig, and dried dates	,,	0.25	0.75
Spices	Microbial decontamination	6.0	14.0

How Irradiation Brings Lethality

On exposing to radiation the water molecules undergoes radiolysis resulting in the formation of free radicals along the path of primary electrons.

$$3\ H_2O \text{———Radiolyses to—–} H + OH + H_2O + H_2$$

These free radicals found along the path escapes as free radicals which bombard with other macro molecules like protein, carbohydrate and fat etc thus altering the organized structure of proteins, carbohydrates, fat and disorient the molecular structure and denatures the enzyme configuration, thus causing lethality to cells of the micro organisms. 1 KGy of radiation breaks 6-10 million bonds on carbohydrate, protein and fat.

Skin Coatings

Waxes

Waxes are generally used for coating fruits and vegetables to improve the appearance of produce or to delay deterioration. Waxes are esters of higher fatty acids with monohydric alcohols and hydrocarbons and some free fatty acids.

Both vegetable and mineral waxes are used in varying composition to make wax emulsions. Waxes are melted, emulsifiers are added and with addition of hot water, oil water emulsions are prepared with continuous agitation. Wax emulsion has been developed to give a protective coating on the skin of fruits and vegetables to reduce transpiration losses, rate of respiration, and prolong the storage life.

Waxing of fruits and vegetables was practiced as early as 12th and 13th century in China. Dipping fruits in molten wax imparted thick coating to the fruits and lead to undesirable changes like fermentation. Hence later improvements were made and a mixture of mineral oil and paraffin wax was applied to the apples. This was further modified and experiments conducted with different vegetable waxes and with water emulsions which gave interesting results regarding reduction in physiological losses in weight, internal atmosphere, respiration rate, decay, shrivelling, scald injury, palatability, taste and flavour in waxed apples during storage. Waxing of fresh fruits and vegetables have assumed a commercial importance in USA, Australia and South Africa.

Tall prolong, a commercial coating that consists of sucrose, esters of fatty acids and carboxy methyl cellulose, could delay the ripening of bananas. The effect is due to the restriction in gas exchange between the fruit and its surrounding atmosphere. This causes build up of carbon dioxide and depletion of oxygen, thus causing an effect achieved as in controlled atmosphere storage. Banana crowns coated with "sempfresh" also showed delayed development of crown rot caused by infection with *Colletotrichum musae*.

CFTRI has developed formulations of wax emulsion Waxol–0-12 and Waxol – W-12 each containing 12 per cent solids for extending the storage life of fresh fruits and vegetables, both at ambient and low temperatures.

The wax emulsions without fungicide in the required concentration did not protect fruit and vegetable against microbial spoilage. Hence permissible concentration of fungicides should be incorporated in wax emulsions prior to its application or may be given as wash treatment prior to skin coating.

All fruits and vegetables except a few do not respond to wax coating. They are cabbage, cauliflower, beetroot, spinach, radish, green peas and French beans.

The concentration of wax emulsion required to obtain maximum benefits vary from fruit to fruit depending up on the skin structure. The approved fungicides incorporated in wax emulsion are 500ppm of 0.25-2 per cent Sodiumhypochlorite, 500ppm of 0.1-1 per cent Benlate, 500ppm of 0.1-2 per cent Thiobendazole.

Fruits and vegetables should be dry, prior to waxing and are waxed by dipping in wax emulsion for 30 to 60 seconds, removed, allowed to drain and dry with the help of a fan as hot air blower or under the shade of a tree by breeze. The fruits prior to dipping should be free of dust, dirt and any microbial infection, emulsions can be reused and is stable up to one year. Proper application of the wax emulsion to fresh fruits and vegetables will not leave any residue or impart any undesirable odour or flavour, bad appearance or interfere with the natural appearance. On the other hand, it imparts gloss to commodity and improves marketability. Very minute quantity goes on to the surface of the fruit and is inconspicuous and is found harmless. One gallon of wax emulsion was found adequate to treat 6000 to 8000 round sized fruits. Wax coatings on fresh fruits and vegetables reduce physiological loss in water due to dehydration. During storage wax and oil coatings retard desiccation in apple, mango, banana and papaya.

All fruits and vegetables have a natural waxy coating on their surface which conserves water. During various handling operations like wiping of fruits with cloth or wrapping it with paper can cause the abrasion which is sufficient enough to impair the protective action of the natural wax layer. This can increase the rate of transpiration and respiration of fruits. Hence a protective coat on fruits and vegetables is given by application of extra continous or discontinuous film on them. The minute particles of wax emulsion clogs the pores of fruit skin [stomatal or lenticels] and continuous film formed after the emulsion dries up, leads to reduction in cuticular transpirational losses of water vapours from fruits. The surface topography and structure of fruit is very important from the view point of coating with waxes.

Studies made on the waxing of fruits in relation to character of protective cover showed that the thickness of the wax film varied widely. Some fruits had areas devoid of wax and there was no consistent relationship between surface tension and coverage. The type of waxes used was important than the thickness of wax coating. The surface character of fruit was important in relation to the kind and concentration of wax used. Solvent waxes were found of lesser value in case of pubescent fruits or fruits having large number of lenticels. The viscosity of the wax solution is to be considered at the time of coating. Hence selection of wax emulsion, composition and concentration to suit the need of a particular fruit is essential. However, the reduction in respiration rate of coated material should not reach critical oxygen entry levels (of less than 3 per cent) in order that the respiration process continues to take place aerobically.

Some attempts have been made to augment existing cuticles and cut down transpiration by applying a wide range of transpiration suppressants such as latex and plastic compounds. Another approach has been made to use salts, such as phenyl mercuric acetate to bring about a premature closing of the stomatal pores. Wax emulsions were thought to partially clog the stomatal pores and thereby reduce the metabolic activities leading to extension of storage life.

Later, improvement have been made on this wax formulation and also anti transpirants like polyethylene emulsion, vapour guard, wilt proof chemicals were developed. Different formulations of wax emulsion by altering the proportion of different vegetables and petroleum waxes showed improvement in the prolongation of shelf life of apples, oranges, green chilies, tomatoes by improving the air permeability and reducing the water vapour transmission rate as observed by in vitro studies on coated craft paper and in vivo efficacies on control of shrivelling in fruits and vegetables.

Inducing or Enhancing Ripening Agents

The rate of ripening can be enhanced or induced by smoking or by the application of ethylene or growth regulators. The use of smoking and ethylene has been in practice to induce uniform ripening of bananas. Application of growth regulators has also been found to accelerate ripening in apple and apricot by 2,4,5-trichlorophenoxy acetic acid (2,4,5-T),in banana by 2,4-dichlorophenoxy acetic acid (2,4-D) in varieties of apple by naphthalene acetic acid (NAA). As a post harvest treatment, 1000-ppm ethephon promotes ripening of tomato, banana,mango etc.

Retarding or Delaying Ripening Agents

Besides physical factors like low temperature, higher concentrations of co_2, low o_2 concentration, waxing, retarding of ripening of fruits can also be achieved by the effect of growth inhibitors like maleic hydroxide (MH) and methyl esters of naphthalene acetic acid(MENA). NAA and 2,4-D, at appropriate dose have successfully retarded ripening and maturation of pineapple and beans respectively.

Maleic hydrazide inhibited sprouting in onions, radish, sugar beets, turnips, carrots and potatoes at a concentration of 2500 ppm, when applied 10-14 days before harvest. In potatoes a range of chemicals have been shown to suppress sprouting by post harvest application for example, 3-chloro-iso-propyl phenyl carbonate.

Pre-cooling

It is the rapid cooling process administered in commodities for removal of field heat. It has to be done immediately after harvest. Pre-cooling reduces the rate of respiration of the freshly harvested commodity accompanied by reduction in metabolic activity and thereby extends the shelf life of the commodity.

Pre-cooling Reduces

1. Field heat
2. Rate of respiration
3. Rate of ripening
4. Loss of moisture
5. Production of ethylene

Pre-cooling method is selected based on

1. Nature of the product
2. Value and quantity of the product
3. Cost of labour
4. Equipment and materials
5. Heat transfer characteristics of the product
6. Surface to volume ratio.

Success of pre-cooling depends on

1. Time between harvest and pre-cooling
2. Initial and final product temperature
3. Accessibility of cooling medium and the produce

Types of Pre-cooling

Room Cooling

This method simply involves placing the crop into a cold store. This may be the same cold store where the crop is to be stored for long periods. The type of room used may vary, but generally consists of a refrigeration unit across which cold air is passed from fan. The length of time required is the only draw back in this method.

Water Cooling

It can be done by flooding, spraying or immersion. Flood system is the most efficient, in which a film of cold water flows rapidly and uniformly over the surface of warm substances. The advantages of this method are speed, uniform cooling and no weight loss or dehydration.

Figure 31: Water Cooling Chamber
Vegetables immersed in ice water for pre-cooling

Vacuum Cooling

Water in the product acts as refrigerant under low pressure obtained in the vacuum chamber. A pressure of 4.6 mm of mercury reduces the boiling point of water from 100°C to 0°C. As the water leaves the product in the form of vapour, the product is cooled by supplying the latent heat required to evaporate water.

Forced Air Cooling

The simplest design is achieved by building parallel stacks of palletized cartons in a refrigerated cold room. The

Figure 32: Forced Air Cooling Unit
Fruits packed in ventilated cartons for pre-cooling
fast method but more expensive

gap between the two parallel rows of pallets is closed off with a cover. A small exhaust fan is placed at one end. The exhaust fan removes air from the enclosed space so that the pressure falls.

A forced air pre-cooler with more than half tonne fruits holding capacity at a time has been designed and developed for pre-cooling studies in citrus fruits. Corrugated fibre board container of 50 x 30 x 30 cm outer dimension is found to be the most suitable for packing Nagpur mandarins during pre cooling.

Heat Treatments

Heat treatment is one of the post harvest treatments in certain commodities mainly to destroy the insects and its eggs.

Different Methods of Heat Treatment

1. Vapour heat treatment (VHT)
2. Hot water dips and sprays
3. Hot air treatment

Vapour heat treatment is used for insect control. Hot water treatment is used for fungal control and also for disinfestations of insects. Forced hot air treatment is used to control both fungal and insect attack and to study the response of commodities to high temperature.

Beneficial Responses of Heat Treatment

1. Slowing of ripening of climacteric fruit and vegetables.
2. Sweetening of commodities either by increasing sugars or by decreasing acidity.
3. Prevention of storage disorders like superficial scald on apples and chilling injury on subtropical fruits and vegetables.

Vapour Heat Treatment (VHT)

It is a method of heating fruits with warm air saturated with water vapour at temperatures between 40 and 50°C to kill insect eggs and larvae as a quarantine treatment before fresh shipment. VHT is widely accepted for disinfestations against fruit flies and other storage rots. The large scale VHT tests have been carried out on Alphonso,

Amrapali, Banganapalli, Chausa, Desheri, Kesar, Langra, Neelam, Suvarnarekha and Totapuri mangoes in India.

VHT consists of a period of warming (approach time) which can be faster or slower depending on commodity's sensitivity to high temperature and the warming period is followed by a holding period when the interior temperature of the produce reaches the desired temperature for the length of time required to kill the insect. Lastly a cooling period which may be air cooling, (slow) or hydro cooling (fast).

Hot Water Dips and Sprays

Hot water dips are generally used for the control of fungal pathogens and spores and latent infections. It is used for the control of infection either on the surface or in the first few cell layers under the peel of the fruit or vegetable. Post harvest dips to control decay are often applied for only a few minutes and temperature used are higher than those for hot air or vapour heat because only the surface of the commodity is heated. Many fruits and vegetables tolerate hot water temperature of 50 to 60°C up to 10 minutes but shorter exposure at these temperatures can control many post harvest plant pathogens.

Low concentrations of fungicides can be applied along with hot water treatment, thus allowing more effective control with a reduction in chemicals. Compounds Generally Recognized As Safe (GRAS) can be applied in hot water to improve the efficiency of their antifungal action. Heated solutions (45°C) of SO_2, ethanol and sodium carbonate can be used to control green mould (*Penicillium digitatum)* on citrus fruits.

A recent innovation in hot water treatment is the development of a hot water spray machine.

This is a technique designed to be part of the sorting line, whereby the commodity is moved by means of brush rollers through a pressurized spray of hot water. By varying the speed of the brushes and the number of nozzles spraying the water, the commodity can be exposed to high temperature for 10 to 60 seconds. The water can be recycled, since the temperature of the recycled water is raised to 50 to 70°C. Organisms that are washed off from the product in the previous washing in the water do not survive. This type of machine is currently used in Israel both to clean and to reduce pathogen population on a number of fruits and vegetables such as mangoes.

Hot water dips have also been tested for their efficiency in disinfesting insects. Hot water is a more efficient heat transfer medium than hot air, and when it is properly circulated through a load of fruits, uniform temperature profile is established. For disinfestations, a longer treatment is necessary than for fungal control because here the total fruit and hot air just around the surface has to be brought to the proper temperature.

Procedure have been developed to disinfest a number of subtropical and tropical fruits including banana, papaya, and mango for various species of fruit fly. In addition, hot water dips are also being investigated for insect control in avocado, citrus, guava, persimmons etc.

The time of immersion can be one hour or more at temperatures below 50°C as against many antifungal treatments that require only minutes at temperatures above 50°C.

For disinfestations of surface insects such as thrips effective time of immersion may be similar to those used for fungal treatments. Some fruits (apricot, nectarine and peach) can be cleared of thrips by 2 or more minutes in water at 48 to 50° C.

The treatment consists of a period of warming (approach time), which can be faster or slower depending on a commodity's sensitivity to high temperature. The warming period is followed by a holding period when the interior temperature of the produce reaches the desired temperature for the length of time required to kill the insects.

The last phase in the cooling period which may be air cooling (slow) or hydro cooling (fast).The different components in the treatment can be manipulated to find the best combinations for elimination of the insect pest without damaging the commodity.

Vapour heat treatments have been developed against the oriental fruit fly in papaya, against the melon fly in mango, against the Caribbean fruit fly in grape fruit and carambola and against codling moth in apple and pear.

Hot Air Treatment

Hot air can be applied by placing fruit or vegetables in a heated chamber with a ventilating fan, or by applying forced hot air, during which the speed of air circulation is precisely controlled. This method heats more slowly than hot water or vapour heat, although forced hot air will heat the produce faster than a regular heating chamber. This method is utilized to study physiological changes in fruits and vegetables in response to heat. Forced hot air is also used to develop quarantine procedures. The reason for choosing this method is that the high humidity in vapour heat may sometimes damage the fruit being treated. The slower heating time and lower humidity with forced hot air causes only less damage. A high temperature forced air quarantine treatment to kill Mediterranean fruit fly, melon fly and oriental fruit fly on papaya also has been developed. This procedure require rapid cooling after the heat treatment to prevent fruit injury as done in citrus.

Hot air can also decrease fungal infections. Heating can reduce the decay caused by *Botrytis cinerea and Penicillium expansum* in apple and *Botrytis cinerea* in tomato. In these cases, long term heating treatments from 12 to 96 hrs at temperatures ranging from 38 to 46°C is used.

References

1. Baldwin, E. A. 1994. Edible coatings for fruits and vegetables: past, present and future. In: Krochta et al. (eds.) Edible coatings and films to improve food quality. Technomic Publishing Co., Lancester.USA.pp 25-64.
2. Barkai, G.R. and Phillips. D.J. 1991. Post harvest heat treatment of fresh fruits and vegetables for decay control. *Plant Dis.*, 75: 1085-1089.
3. Combrink, J.C., Benic, L.B., Lotz, E. and Truter, A.B. 1994. Integrated management of post harvest fruit quality. *Acta Hort.* 368: 657-666.
4. Dalal, V.B., 1989. Pre-treatment, packaging, storage and transportation of fresh fruits and vegetables. In: Trends in Food Science and Technology, Proceedings of IFCON, 1988. AFST, India, Mysore, pp. 334-344.

5. Dalal,V.B., Eipeson, W.E. and Singh, N.S. 1971. Wax emulsion for fresh fruits and vegetables to extend their storage life. *Ind. Food Packer,* 25: 9-15.

6. Dasgupta, M.K. and Mandal, N.C.1989. Post Harvest Pathology of Perishables. Oxford and IBH publishing Co. New Delhi, P. 2-6

7. Drake, R. S. and Nelson, J. W. 1990. Storage quality of waxed and non waxed delicious and golden delicious apples. *J. Fd Qual.* 14: 331–341.

8. Habibunnisa, 1993. Pre and post-harvest treatments of fruits vegetables. Paper presented in the Advanced Technology Training Programme, 27 September 1993, CFTRI, Mysore.

9. Krishnaprkash, M.S. 1993. Pre-cooling of fruits and vegetables. Paper presented in the Advanced Technology Training Programme, 27 September 1993, CFTRI, Mysore.

10. Lurie, S. 1999. Post harvest treatments for Horticultural crops. *Hort. Rev.* 22: 91-121.

11. Merino S. R., Eugenio M. M., Ramas. A. U. and Htrnandez, S. U. 1985. *Fruitfly Disinfestation of Mangoes (Mangifera indica* var. 'Manila Super') by *Vapour Heat Treatment.* Ministry of Agriculture and Food, Bureau of Plant Industry, Manila, Philippines

12. Rij, R. 1979. Handling, precooling and temperature management of cut flower crops for truck transportation. USDA Science and Education Administration, AAT-W-5, Leaflet 21058.

13. Urso, C. 1987. Procedure and machine for coating of fruits and vegetables with plastic film. European patent E.P.O. 156718 B 1 (1987) France.

14. Venugopal, V and Warrier, S. 2001. Radiation processing and value addition. Library and information service division publication, BARC, Mumbai.

Chapter 6

Packaging

Packaging is becoming an essential part of the value chain analysis regarding food safety, organo–leptic characteristics, agro-economics and flexibility. Packaging is also of great importance in the final choice of the consumer because it directly involves convenience, appeal, information and branding.

Fresh produce properly treated, sorted, graded needs proper package for protection during transportation and storage. Methods of packaging can affect the stability of products in the container during shipping depending upon the extent of the container protecting the product. For example, delicate and high priced products are packed in trays or in fibre board boxes, whereas other products are simply put directly in the boxes together.

Pre-packaging or consumer packaging is also convenient for retailers as well as customers and therefore adds value to the produce.

It generally provides additional protection for the products. At the same time, cover of non-biodegradable plastic trays and wrapping materials often seen in modern supermarkets, creates an extra burden of waste disposal and damages the environment.

The key functions of packaging are :

1. To assemble the produce in convenient units for handling
2. To protect the produce during handling, transportation, storage and marketing

Universally Standardized Requirements for Packaging

(*a*) The package must have sufficient mechanical strength to protect thc content during handling, transportation and stacking one over the other.

(*b*) The package must meet handling and marketing requirements in terms of size, shape and weight in accordance with International standards. The current trend is to avoid too many sizes and shapes of packages. Palletisation and mechanical handling makes standardization essential for economic operation.

(*c*) The material of the package must not contain any toxic chemicals which could transfer or produce toxin in to human beings.

(*d*) The package should allow rapid cooling of the contents

(*e*) The package should be stable to moisture and high humidity.

(*f*) The packages should be stackable and interlock able.

(*g*) The packages should be re-usable or recyclable for easy disposability.

(*h*) The package should provide adequate ventilation

(*i*) The package should be of suitable standard depending on the market demand.

(*j*) It should be cost effective in relation to market value of the commodity.

Advantages of Packaging

Packaging provides a beneficial modified micro environment that helps in:

1. Minimizing post harvest loss by protecting against mechanical damage, microbes, pests, dust and air pollution, moisture loss, pilferage etc.
2. Efficient handling and marketing
3. Giving better appeal so as to promote sale.
4. Providing hygienic condition within the package.
5. Enhancing marketable distance and time.
6. Protecting nutritive quality
7. Preventing contamination by other commodities
8. Providing information about the contents
9. Ready to use facility

Classification of Packaging Materials

Traditional Ones

1. **Natural materials**: Packaging containers made of bamboo, straw, palm leaves etc

2. Natural and synthetic fibre–sacks made of jute, cotton, woven plastic and paper.
3. Wood: wooden crates, wire bound veneer and crates.

Recent Ones

1. Corrugated fibre board: ventilated and non ventilated
2. Plastic crates
3. Molded trays, paper pulp and plastic
4. Net/Mesh bags/Sleeve pack
5. Plastic films/bags/boxes

Specialized Ones

1. Cling film
2. Shrink wrap film/Stretch film
3. Flexible packaging materials for modified atmosphere packaging : Low density polyethylene (LDPE), Cellophane, Rubber hydrochloride, Polyvinyl chloride (PVC)
4. Fancy packaging materials

Natural Materials

This type of packages includes baskets and other traditional containers made from bamboo, palm leaves, straw etc.

The characteristic features of such containers are:

Labour cost and raw material cost involved in making such containers are low and they provide good ventilation.

Figure 33: Bamboo Baskets with Mangoes

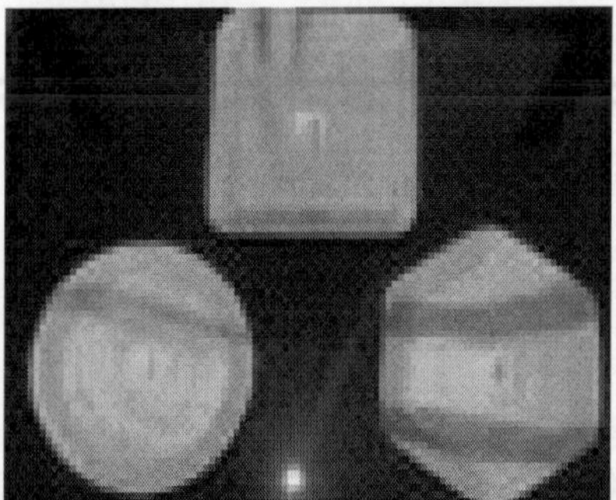

Figure 34: Arecanut Leaves Sheath Trays

Drawbacks

1. Difficult to clean if contaminated with decaying organisms
2. Lack rigidity and bend out of shape when stacked

3. Load badly because of shape
4. Causes pressure damage when tightly filled
5. Sharp edges or splinters cause cut and puncture damage to the contents.
6. Less life, so should be replaced frequently.

Natural/Synthetic Fibres

Sacks-These types of packaging materials are made from jute or cotton or woven plastic or paper.

These are inexpensive and readily available; moreover they are reusable and have good load bearing capacity.

Figure 35: Jute Bags with Grains

Drawbacks

1. Lack rigidity and handling can damage contents
2. Too large for careful handling; if dropped it bruises the contents.
3. Impair ventilation if stacked
4. Difficult to stack on pallets

With these drawbacks also these are widely being used in Kerala for bulk packaging and transportation of fruits and vegetables. So a modification suggested is to place a hollow bamboo pole with holes made on it at regular intervals having about one meter long with lesser diameter to be inserted into the centre of the sack and fill the sack with respiring commodities so that the heat generated while respiring can be vented through this bamboo right from the bottom of the sack holding about 100 kg materials.

Wood

Soft wood planks can be used to assemble into wooden boxes of required dimensions known generally as wooden crates. If it requires more cushioning property this could be woven outside using coir rope then it is known as wire bound veneers.

Advantages in terms of characters like:

1. Rigid and reusable
2. Facilitates ventilation
3. Stack well on trucks if made to a standard size.

Drawbacks

1. Difficult to clean adequately for multiple use

Figure 36: Wooden Boxes with Tomatoes

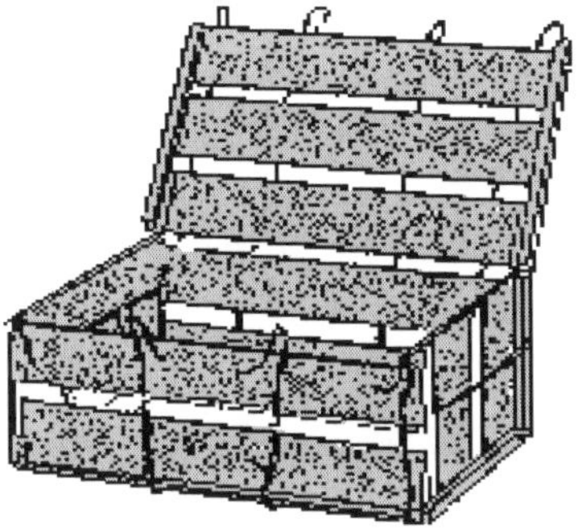

Figure 37: Wire Bound Veneers

2. Heavy and costly to transport
3. Sharp edges, splinters and protruding nails damage contents

Corrugated Fibre Board (CFB) Boxes/Cartons/Cases

Also known as cardboard or fibre board or pasteboard boxes. It is made from a layer of corrugated fibre board sandwiched between two additional layers of fibre board.

The characteristic features are

1. Most widely used and acceptable packaging material.
2. Lightweight (20 to 25 percent lighter when compared to wooden box of similar size) and clean.
3. Excellent cushioning property and smooth nonabrasive surface.
4. Low cost, reusable and recyclable.
5. Excellent printability.
6. Easy to set up for storage.

Figure 38: Corrugated Fibre Board Box with Vegetables

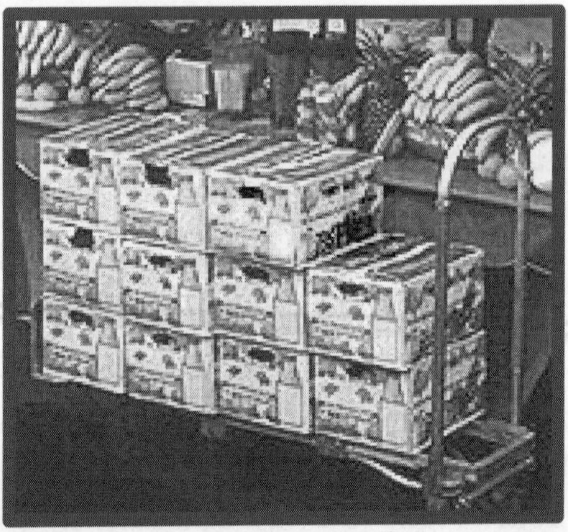

Figure 39: Waxed Chip Board Cartons with Fruits

7. Available in wide range of sizes, designs and strengths.

Drawbacks

1. Easily damaged by careless handling and stacking.
2. Weakens on exposure to moisture.

Plastic Crates

These are mostly moulded from high density polyethylene (HDPE). This packaging material has the following features:

1. Reusable, strong, rigid and smooth
2. Easy to clean
3. Good ventilation
4. Can be made to stack when filled and nest when empty, so space saving.

Figure 40: Plastic Crates with Potatoes

Drawbacks

1. Costly
2. Mostly imported adding to the cost
3. Not foldable
4. Deteriorate when exposed to sunlight unless treated with UV inhibitor, which adds to cost.
5. Though costly, capacity for reusage make it an economical package.

Moulded Trays

It may be moulded from paper pulp or plastics. It is suitable for packaging individual fruits and vegetables.

Paper Pulp Moulded Tray

Made from recycled paper with starch binder therefore it is less expensive.

Absorbs surface moisture from the product so it is good for small fruits and berries which are easily damaged by water.

Biodegradable and recyclable

Figure 41: Moulded Trays with Apple

Thermoformed Plastic trays

These are trays made from thermoplastics which can be softened by heating and hardened by cooling. They can be regarded as derivatives of ethylene. They are also known as vinyl plastics, which are containing chlorine or fluorine or phenyl groups.

Advantages

1. Rigid packaging
2. Immobilization of the produce within the pack

Figure 42: Thermoformed Plastic Trays with Fruits

3. Suitable for microwave cooking without grease and loss of vitamins.

Method of Packaging

The produce is filled in trays, over wrapped with a heat shrinkable film and passed through a hot tunnel. The most commonly used micro-ovenable packages are crystallized polyester (CPET) in the form of trays. They are flexible in shape and design and resistant to oil and grease. Polypropylene co-extruded with barrier resins like Ethyl Vinyl Alcohol (EVA) is used when a longer shelf life is required.

Plastic Film Bags

Polyethylene and polypropylene bags are mainly used, which have the following characteristic features:

1. Low cost
2. Widely used for consumer size packs in fruit and vegetable marketing
3. Retain water vapour so as to reduce water loss from the contents.

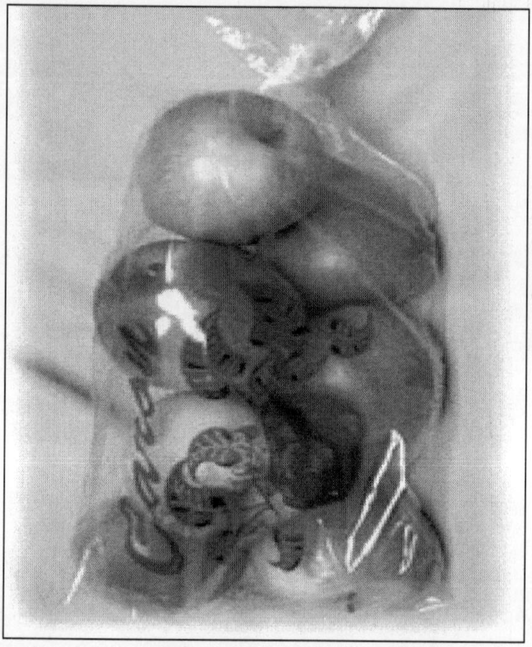

Figure 43: Plastic Bag Containing Apples

Drawbacks

1. No protection from injury caused by careless handling
2. Heavy buildup of condensation may lead to decay
3. Rapid build up of heat if exposed to sunlight
4. Permit only slow gas exchange. Vapour and heat along consumer packs of plastics are not recommended in tropics except in stores with refrigerated display cabinet.

Plastic Boxes

Ideal for consumer packs and has good sales appeal.

Figure 44: Plastic Box with Okra

They are rigid containers most suited for packaging soft and delicate commodities.

Can be fabricated into different shapes and sizes to suit the requirements.

Net or Mesh Bags

Widely used for packing fruits like apple, citrus, guava, sapota etc. and have the following features:

1. Sturdy
2. Low cost
3. Uninhibited airflow
4. Attractive display which stimulates purchase.

Drawbacks

1. Large bags do not palletize well whereas small ones do not efficiently fill the space inside CFB boxes.
2. Do not offer protection from rough handling
3. Little protection from heat and contaminants

Sleeve Packs

Heat shrinkable film of 1.5 to 2 mm thickness is used. The following are the features of this type of packaging.

1. Immobilization of packed fruits
2. Superior visibility that gives a good sales appeal
3. Low cost
4. Better protective qualities

Cling Film

Polyethylene film of 15 microns thickness is used. Features that make it an ideal packaging material are:

1. Low water vapour transmission rate
2. High gas permeability
3. Intimate package with the individual produce
4. Keeps the produce fresh, dust and insect free
5. Self-sealing types are available commercially.

Shrink Film or Stretch Film

Principle involved is plastic film like polyethylene, poly styrene, polyvinyl chloride, polyester, rubber hydrochloride

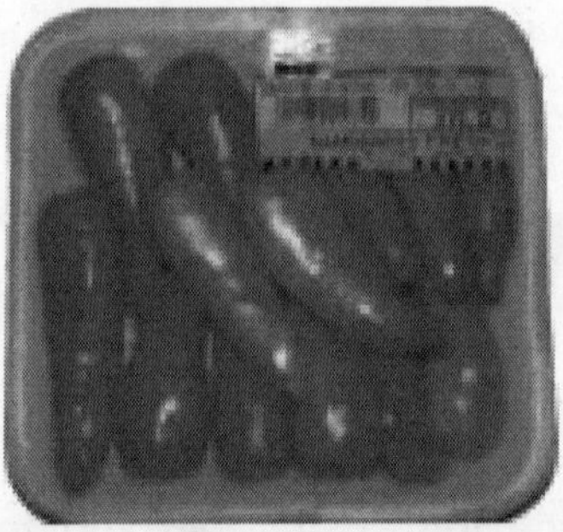

Figure 45: Carrot Individually Wrapped with Cling Film

Figure 46: Shrink Wrapping Machine

etc has heat shrinkable nature. By stretching the film under controlled temperature and tension, the film which is wrapped over the produce stretches and then contract by cooling.

The characteristic features are

1. Can be used as over wraps on individual fruits and over consumer size trays or pack
2. Shrink on exposure to moderate temperature
3. Heat supplied by electric resistance coil is just enough to shrink the film but not enough to harm the produce.
4. Mineral impregnated polyethylene film adsorbs and removes ethylene gas and has excellent permeability and good deodorizing properties.
5. The addition of anti-fogging treatment to the film reduces the formation of water drops and the protection for mould and bacterial growth.

Benefits of Shrink Wrapping

1. Prolongs the life, freshness and colour of packaged fruits, vegetables and cut flowers.

2. Possible to store and ship ethylene generators and ethylene sensitive products in proximity to each other.
3. The film maintains high humidity levels reducing water loss from packaged produce.
4. The potential for mould and bacterial growth and spoilage is reduced by anti-fogging treatment.
5. A good surface for stick on labels
6. Protects the produce from diseases
7. Reduce mechanical injury
8. Some of very highly perishable commodities has shown to get the shelf life extended by Shrink wrapping; example lettuce keeps good up to 6 weeks and carnation up to 3-4 weeks.

Active Packaging: An Innovative Approach of Food Preservation

Active packaging is a group of technologies in which the package is active and is actively involved with food products or interacts with internal atmosphere to extend shelf life while maintaining quality and safety.

Active packaging, some times referred to as 'interactive' or 'smart' packaging, which is intended to sense internal or external environmental changes and to respond by changing its own properties or attributes.

Potential Technologies Used in Active Packaging

Oxygen Scavenging

Most important objective of active packaging has been the removal of oxygen using various techniques *viz.*, absorption, interception and scavenging. The most widely

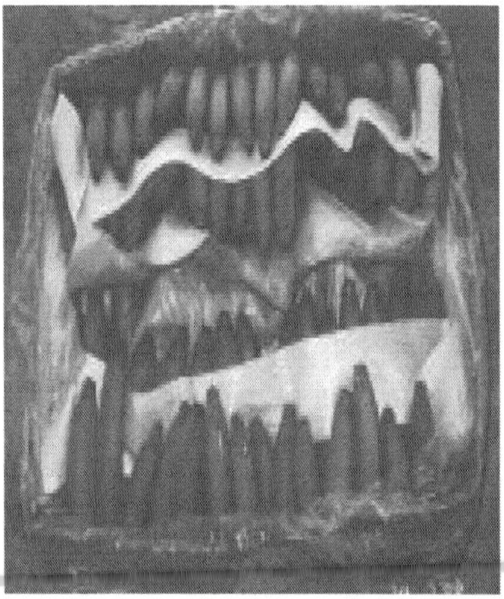

Figure 47: Active Packaged Banana Hands with Sachets Placed in the Centre Containing the Active Agents

researched and patented area of active packaging is the use of oxygen absorbing systems. Mitsubishi Gas Chemical (MGC) company was the first to launch ferrous based oxygen scavenger sachets under the trade name 'Ageless' and till date they are regarded as the pioneer in the oxygen scavenging technology.

Antimicrobial Packaging

Active packaging system comprising incorporation of antimicrobial agents into polymer surface coatings and surface attachments can be of immense value. Most widely publicized antimicrobial agents are silver salts on zeolite incorporated in plastic films and sheets or on material surfaces and into absorbent pads for fresh meal and produce.

Ethylene Control

Ethylene is a hormone, which accelerates ripening in

fruits followed by senescence. Removal of ethylene from plant environment can significantly retard post harvest catabolic activity in fresh produces and modified atmosphere preservation process. 'Frisspack' is a paper incorporated with $KNnO_4$ for use in corrugated fiberboard cases. Activated charcoal impregnated with a palladium catalyst and placed in proper sachets effectively removes ethylene by oxidation from packages of minimally processed kiwi, banana, broccoli and spinach.

Moisture Control

Since the metabolism of fats and carbohydrates produces water, condensation or sweating is a problem in many kinds of packaged food, particularly fresh fruits and vegetables or minimally processed and prepared foods. Use of humectants between two highly permeable layers of a plastic film has been found to buffer the humidity inside the food package.

In Package Fumigant

1. Slow release system developed by CFTRI consists of a mixture of 99g KMS and one gram Citric Acid (CA). Approximately 500±5mg of mixture is packed in craft paper bag of 4" x 2½" and folded and stapled. Eight such unit packets (4 g of material) were used for 4 kg of grapes.
2. Combination of slow and quick release system contains an equal proportion of a mixture of 100 g KMS and 60g CA which releases SO_2 quickly. This gave significantly better protection to grapes without adversely affecting quality during transportation and controlled spoilage in conventional packaging.

Gas Permeability Control

O_2 Control

Package material designers have developed high oxygen permeable packaging material. O_2 permeation rates of 18.06 in %m^2/day and CO_2 permeation rates of 7.74 in %m^2/day are achievable and effective at various controllable ratios for high respiration rate for fresh produces such as broccoli, mushroom, asparagus and strawberry.

CO_2 Control

It is a complementary approach to O_2 control. A mixture of iron powder and Ca $(OH)_2$ in the form of sachet, which scavenges both $O_2$and CO_2, has been used to package fresh ground coffee in flexible bags.

Active packaging is bound to emerge as a concrete step in the evolution process of advanced packaging technologies to meet the requirements of modern innovative food processing.

References

1. Abe, K. and Watada, A. F. 1991. Ethylene Absorbent to maintain quality of highly processed fruits and vegetables. *J. Fd Sci.* 56: 1589-1592.
2. Anand, J.C. and Maini, S. B. 1982. Fibreboard packaging for fruits. *Ind. Hort.*, pp 31-55
3. Banks, N.H. 1984. Some effects of Tal Prolong coating on ripening of bananas. *J. Exptl. Botany* 35: pp 127-137.
4. Brody, A. L., Strupinsky, E. R. and Kline, L. R. 2001. Active Packaging for Food Applications. Technomic Publishing Co., Inc. Lancaster, USA.

5. Floros. J. D., Dock, L. L. and Hall, J. H. 1997. Active Packaging Technologies and Applications. *Fd. Cosmetics and Drugs Packaging.* 20: 10-17.

6. Joshi, G.D. and Roy, S.K. 1984. Standardization of packaging of fresh mango fruits (cultivar Alphonso) for transportation and storage. *Proceedings of the National conference on packaging of fresh and processed foods.* March 2 and 3 1984, Calcutta.

7. Maini, S.B Lal and Anand, J.C. 1993. Fruit packaging. Advances in Horticulture. (Ed.) Chadha, K.L. and Pareek, O.P., Malhotra Publishing House, New Delhi. (4): p. 1775.

8. Peleg, K. 1985. Produce handling packaging and distribution. AVI publishing company, Connecticut. pp 233-236.

9. Rooney M. L. 1995. Overview of Active Food Packaging. In Active Packaging (Ed. Rooney, M. L.). Blackie Academic and Professionals Glasgow, UK, pp. 120-123.

10. Satyan, S.H., Scott, K.J., Graham, D. 1992. Storage of banana bunches in sealed polyethylene tubes. *J. Hort. Sci.* 67: 283-287.

11. Shetty, K.K., Klowden, M.J., Jang, E.B., Koshan, W. 1989. Individual shrink wrapping technique for fruit fly disinfestations in tropical fruits. *Hort. Science* 24:3 17-319.

Chapter 7

Storage

Many horticultural crops have a relatively short harvesting season. Fruits and vegetables are living organisms. Their condition and marketable life are affected by many things as temperature, humidity, composition of the atmosphere which surrounds them, level of damage that has been inflicted on them and the type and degree of infection with micro organisms.

Purpose of Storage

1. Minimize post harvest loss
2. To tie over the surplus production
3. Make it available where it is not available
4. Stabilize the market
5. Make it available during off season

Storage Methods

1. Cold storage (CS)

2. Controlled atmosphere storage (CAS)
3. Modified atmosphere storage (MAS)
4. Modified atmosphere package (MAP)
5. Hypobaric storage (HS)
6. Evaporative cool storage (ECS)

Fruits and vegetables acclaimed universally as protective foods are the important constituents of our daily diet. These protective foods are highly perishable after their harvest and hence their preservation involves maintenance of fruit or vegetable tissue in physiologically sound and fresh condition. In our country the wastage of fruits and vegetables are estimated to be about 20 to 30 per cent during harvesting, handling, transportation and storage.

In order to prevent losses in storage of fresh produce, one of the treatments is to reduce the temperature of the perishable commodity immediately after harvest, which accomplishes considerable fall in the rate of respiration which reduces the built up of respiration heat, thermal decomposition and microbial spoilage and also retention of quality as near the fresh condition as possible with subsequent increase in their storage life without adversely affecting their commercial or nutritive value for normal processes of ripening.

Cold Storage

Fruits and vegetables after separation from the parent plant (after harvest) continue to respire(live) at the expenses of the stored food material which depending upon the rate of metabolism, declines and accelerates the process of ageing as evidenced by shriveling accompanied by changes in

Figure 48: Fruits and Vegetables in Cold Room

texture, flavour and quality. Fruits and vegetables respire by taking in oxygen and giving out carbon dioxide and heat and moisture loss through natural pores in the skin by transpiration. The enzymatic breakdown of reserve food material and consequent loss of resistance to microbial and fungal invasion drastically cuts down the storage life of fresh fruits and vegetables at room as well as low (refrigerated) temperatures. Removal of heat from freshly harvested commodity without loss of quality is the most desired method for extension of storage life of perishables. Pre-cooling and refrigerated storage at optimum conditions reduce the metabolic rate with relative increase in their storage life. The storage life of freshly harvested commodity is governed by the following factors:

1. Prevailing climatic conditions and cultural practices of the region
2. Variety and storage of maturity at harvest
3. Incidence of infection at the time of harvest.

4. Method of picking, handling and the extent of damage prior to storage
5. Time lag between harvest of the commodity and its subsequent storage
6. Rate of cooling
7. Storage temperature and relative humidity
8. Rate of accumulation of CO_2 concentration inside the refrigerated storage chamber.
9. Air distribution system in the refrigerated cold room
10. Post harvest treatments, size of package, quantity per package, ventilation/air circulation inside refrigerated storage.
11. Sanitary condition inside the refrigerated storage rooms.

It is one of the widely adopted method for bulk handling of the perishables between production and marketing or processing. Maintaining adequately low temperature is critical, as otherwise it will cause chilling injury to the produce. Also relative humidity of the store room should be kept as high as 80-90 per cent for most of the perishables, below (or) above which has detrimental effect on the keeping quality of the produce.

Principle of Cooling

Transfer of heat occurs in three ways, *viz.*, conduction, convection and radiation. During cooling of fruit or vegetable, heat moves from the interior to the surface, principally by conduction. If there are air voids around the seed or in the core area or in the space between the cells the heat would be transferred by convection. The cooling medium whether water or air removes heat from the surface of the fruit by convection and then transfer the heat from the surface to the refrigeration system.

To cool a produce depends on its specific heat, its initial and final temperature, heat of respiration etc. Since fruits and vegetables contain around 80 per cent water, their specific heat are almost nearer to that of water. However, the thermal conductivity differs from fruit to fruit and hence resistance to heat conductivity differs.

The cooling curve is generally exponential and nearly so when the temperature of the cooling medium is constant and the temperature of the gradient has been established. A time constant can be introduced and this has been designated with the value 'Z' or half cooling time. It is the time taken for produce temperature to be reduced to half the initial difference in temperature between produce and cooling medium. It is based on the characteristics of produce, the package and independent of actual temperature. In simple cooling rate the temperature gradient between product and coolant changes. Thus, heat is removed very rapidly during first half cooling period where the difference between product and coolant is maximum. Heat removal then becomes progressively slower during subsequent half cooling period and eventually becomes nil as the temperature of the product and the set temperature (optimum low temperature) is achieved.

Mechanism of Heat Removal by the Refrigeration System

By both first and second order laws of thermo dynamics, the refrigerant gas (Freon) takes up energy as it expands from compressed liquid state to its original gaseous state. This heat is taken up from the atmosphere or chamber or environment in which it gets expanded through the expansion value in the refrigeration system where the commodities for cooling are loaded. The gas first compressed to liquid state in the compressor of the system and due to the increased pressure the gas will get heated up. To remove this heat it is allowed to get passed to a condenser where the temperature is removed either by air cooling or water cooling, thus the liquid refrigerant under pressure devoid of heat will be received in to a receiver and from there to expansion chamber. The liquid Freon will tend to go back to gaseous state in the expansion chamber (Evaporator) for which heat will be taken up from the commodities and thus the commodities get cooled. The cycle will continue till the set temperature is achieved.

Table 11: Optimum Storage Life at Optimum Temperatures for Various Important Fruits and Vegetables

Commodity	*Optimum Storage Temperature (°C)*	*Optimum Storage Life (Weeks)*
Fruits		
Apple	2-3	16-17
Banana	12-13	3-4
Grapes	0-2	5-7
Guava	8-10	2-5
Jackfruit	11-13	7-8
Limes	11-13	8
Mango (green)	11-13	4
Oranges	11-13	8
Sapota	19-21	3
Tomato (mature breaker stage)	16-20	3
Pineapple	8-10	6

Contd...

Table 11–Contd...

Commodity	Optimum Storage Temperature (°C)	Optimum Storage Life (Weeks)
Vegetables		
Beans	0-2	3
Beetroot	0-2	7
Brinjal	8-10	4
Cabbage	0-2	12
Cauliflower	0-2	7
Capsicum	11-13	3
Cucumber	7-8	2
Ginger	2-3	14
Indian goose berry	2-3	8
Nhol-khol	0-2	12
Lady's finger	8-10	2
Onions	0-2	24
Potato	2-3	34
Snake gourd	18-21	2
Tapioca	0-2	23

Storage Disorders and Chilling Injury

Storage disorders include all the physiological disorders. Different fruits and vegetables below its optimum storage temperature is susceptible to various kinds of disorders including fungal infections. Many tropical and subtropical varieties of fruits and vegetables when stored below 50°F (10°C) for a longer time suffer physical and physiological injuries. These injuries are of the following types.

1. Superficial scald
2. Carbon dioxide injury

Table 12: Treatment of Fruits and Vegetables to Control Spoilage During Refrigerated Cold Storage and Transit

Commodity	*Physical Method*	*Chemical Method*
Apple	Pre-cooling in air, cold storage 0-2°C	Washing and dipping in fungicides (500 ppm) wrap in chemically (diphenyl) treated papers/skin coating with wax emulsion
Banana (mature green)	Hydro–cooling, pre-cooling in cold air within 12 hrs and cold storage at 13°C	Washing in fungicide (500 ppm), application of antifungal paste, skin coating with wax emulsion.
Citrus	Pre-cooling in air or hydro-cooling and then cold storage at 13°C	Washing in fungicide, (500 ppm) skin coating with wax emulsion, wrap in chemically (diphenyl) treated paper
Grapes	Pre-cooling in air and cold storage at 0-2°C	Fumigation with SO_2 @1000 ppm in package fumigant packets of KMS (Potassium metabisulphite)
Mango (mature green)	Hydro-cooling or pre-cooling in air and cold storage at 13°C	Washing in fungicide (500 ppm) fumigation with ethylene oxide, skin coating with wax emulsion
Potato	Pre-cooling in air and storage at 0-2°C	Washing in fungicide (500 ppm) and treat with sprout inhibitors such as maleic hydrazide.
Tomato (Breaker pink)	Pre-cooling in air or hydro-cooling and cold storage at 13 °C	Washing in fungicide (500 ppm)/skin coating with wax emulsion
Green vegetables	Hydro-cooling and cold storage	Washing in fungicide (500 ppm) prior to storage

3. Core flesh
4. Breakdown of the flesh of the stored commodity:
   (*a*) Low temperature breakdown (LTB)
   (*b*) Senescence breakdown
5. Water core
6. Bitter pit
7. Freezing injury
8. Chilling injury

Superficial Scald

The common feature of this is that areas of the skin of the stored commodity turns brown. These areas are very slightly sunken and the lenticels stand out as raised uninjured spots, *e.g.* apples, pears, peaches.

Carbon Dioxide Injury

Usually occurs in Controlled Atmosphere Storage (CAS) where sometimes carbon dioxide concentration goes higher. The periphery of the internal tissues turns slightly brown initially and as the time of exposure to CO_2 prolongs the tissue turns deep brown; *e.g.* apples, pears, mangoes.

Core Flesh

Core flesh is yellowish pinkish discolouration of the core of the apples. It may appear as a ring of damaged tissue or it may involve the whole area of the core. The incidence of the disorder is aggravated by increased CO_2 concentrations in the core atmosphere. Storage in virtual absence of CO_2 (at the initial stage) and in low concentration of oxygen gives good control of core flesh.

Low Temperature Breakdown (LTB)

Low temperature breakdown of apple is seen in the cortical tissue as a general browning of the flesh which can vary in intensity from season to season. The vascular tissue (of the commodity) is picked out as dark brown specks. As the disorder progress the skin eventually becomes discoloured and apparently water logged giving a dark translucent appearance. The cut surface is usually moist.

Senescence Breakdown

Senescence breakdown of apples and pears is a disorder associated with over maturity. One which is progressive in that, it will develop further at high temperatures after the fruit has been removed from store. It varies in appearance and breakdown and the flesh may eventually becomes 'mealy'. It appears inwards to be worse when viewed externally than when the commodity is cut open.

Water Core

As the name implies is a flesh breakdown which follows the abnormality of the flesh. Water core is a condition in which parts of the flesh of the commodity appear to be translucent and 'glossy' because the intercellular spaces have become injected with the 'sap'. They can appear anywhere in the flesh and may be near to the surface of the skin. Water core disappears more rapidly at higher storage temperatures. However, badly affected fruits cannot recover from water core disorder.

Bitter Pit

In this, small brown dry areas disfigure the flesh. The location of the pit is usually below the skin. But in severe cases the pits may extend right up to the cortex. Under the microscope the pitted areas are seen to consist of dead collapsed cells. Lack of calcium salts causes bitter pits. Pre harvest spray of calcium reduces bitter pit *e.g.*, pears, apples, guavas.

Freezing Injury

Freezing injury occurs in commodity stored below 0°C. The affected fruit externally has an irregular shape caused by tissue collapse. The juice will stream out from the injured cut tissue even under slight pressure. In apples, freezing injury characteristically occurs in cone shaped segments with the apex at the core.

Chilling Injury (CI)

Chilling injury is a major problem in post-harvest handling of fruits and vegetables because many tropical and sub-tropical fruits and vegetables are sensitive to low temperature storage below 10°C and develop symptoms of chilling injury making the commodity unfit for marketing.

It is a phenomenon during which the symptoms like skin discolouration and browning, pitting of the skin, water soaked spots, soggy flesh and failure to ripen, when the commodity is removed to room temperature (RT). The chilling injury symptoms are visible only after 2 or 3 days storage at room temperature.

Control of Chilling Injury

Temperature Pre-conditioning

Gradual reduction of storage temperature by 1°C of the cold room is found beneficial in alleviation of CI. Pitting in stored banana was found reduced from 90 per cent to 8.9 per cent when stored at temperature of 13°C, 10°C, 8°C at 4 days intervals.

Regulating Humidity

Storage humidity at 95-100 per cent was found to reduce CI symptoms in 'Gros Michel' bananas stored at 10° C when they are covered in polythene bags.

Table 13: Important Commodities Sensitive to Chilling Injury at Low Temperature and the Symptoms Exhibited

Commodity	*Approx. Temp**	*Symptoms Upon Transfer to Room Temperature (RT)*
Banana green	13°C	Browning of the skin, smoky appearance, pitting, browning of the pulp, failure to ripen at room temperature (RT)
Papaya	8°C	Discoloration of skin, pitting, water soaked pits, soggy flesh, pitting, failure to ripen at RT. Off flavour development and poor carotene content.
Mango	8°C	Discoloration and browning of skin, pitting of the skin, smoky appearance, failure to ripen at RT. Poor carotene development in the flesh.
Pineapple	8°C	Brown or dull skin color, water soaked flesh, wilting of the crown, failure to develop full flavour at RT.
Sapota	3°C	Failure to ripen at RT
Tomato green	8°C	Failure to develop red color at RT, pitting of the skin of green tomatoes.
Limes	11-13°C	Pitting and browning of the peel, darkening of the oil glands in the skin, browning of the flesh
Capsicum	11°C	Pitting of the skin, soggy spots in the skin

* Below which chilling injury starts

Intermittent Warming

Intermittent warming (IW) of the commodity for every 5 days at low temperature and then 2 days at room temperature found effective to control CI. Raw green mature papaya stored at 7°C for 5 days and then transferred to room temperature for 2 days did not exhibit CI symptoms up to 3 weeks storage at 7°C. Similarly green mature bananas responded to this treatment with control of CI symptoms when removed to room temperature after 3 weeks with intermittent warming at 9°C.

Role of Wax Coating in Controlling the Chilling Injury

Suitable concentration of wax coatings of the skin of papaya, banana and mango stored at 8°C for two weeks did not develop CI symptoms when removed to RT.

Role of Modified Atmosphere Storage in Controlling the Chilling Injury

When the fresh commodity is enclosed in the thin low density polyethylene(LDPE) bags and stored at chilling temperature, the commodity stored, resisted CI upto 2 weeks. Banana, papaya and mango when enclosed in sealed polyethylene bags develop modified atmosphere and humidity inside the bags and helps to alleviate CI symptoms. The fruits when removed to RT after 2 weeks of storage at chilling temperature of 8°C, did not show CI symptoms.

Theories of mechanism of CI have been worked out by many research workers which involves changes in membrane permeability, abnormal respiration, changes in cell wall structure, impairment of enzyme activity of the chilled injured plant material.

Controlled Atmosphere Storage(CAS)

CAS is defined as an atmosphere in which the oxygen level is brought down to low level and carbon dioxide level is brought to high level, created by natural respiration of the packed commodity or by artificial means. These gases are controlled by a sequence of measurements and corrections through out the storage period.

CAS involves control of certain gases (O_2 and CO_2) around the immediate surface of the commodity and within the fresh fruits and vegetables; hence there is a reduced respiration rate. The level of these gases are maintained by the constant monitoring and adjustment of the CO_2 and O_2 levels within gas tight stores or containers. Due to respiration the level of gases constantly changes and the gases are measured at intervals and adjusted to the pre determined level by the introduction of fresh gas or nitrogen or passing the store atmosphere through a chemical to remove CO_2.

Table 14: CA Requirements for Some Fruits

Fruit	*Temperature*	*O_2 %*	*CO_2 %*
Banana	13-14	2-5	2-5
Mango	10-14	2-5	5-10
Papaya	7-13	2-5	5-8
Pineapple	7-13	2-5	5-10

Controlled and Modified Atmosphere Storage of Tropical Fruits

The maintenance or improvement of the post harvest quality and the post harvest life of fresh fruits and vegetables is becoming increasingly important. With the annual production of over 49.3 million tonnes of fruits, India has emerged as the second largest producer of fruits in the world. Fruits in general have short shelf life and they loose their orchard freshness shortly after they are harvested from the tree. The post harvest losses due to improper handling, packaging and storage of fruits is very high. About 25-30 per cent of the fruits produced in the country is annually lost. Lowering of fruit temperature can extend the shelf life for a long period but many of the tropical fruits are chilling sensitive. This does not permit their maintenance in low temperature and as a consequence these crops have a relatively short post harvest life compared to many temperate and subtropical fruits. Most tropical fruits have a post harvest life of only few weeks at the most. Controlled and modified atmosphere storage are technologies which can increase the marketable life of many tropical fruits.

Advantages of Controlled and Modified Atmosphere Storage

1. Retardation of maturation, ripening and senescence during storage.
2. Alleviation or control of chilling injury
3. Reduce the incidence of diseases
4. Reduce the incidence of physiological disorders
5. Reduce the incidence of insect attack

Modified atmosphere and controlled atmosphere with an oxygen concentration less than or equal to 1.0 per cent and carbon dioxide concentration more than or equal to 5 per cent have insecticidal and fungi static effects.

Characteristics Ideal for Successful CA Storage of Fruits

1. A long post harvest life
2. Resistance to chilling injury
3. A large range of non-injurious atmosphere

4. Resistance to fungal and bacterial attack
5. Adaptation to a humid atmosphere
6. A climacteric fruit that can be ripened during or after storage
7. Absence of negative controlled atmosphere residual effect
8. Ethylene tolerance

Effect of Low Levels of Oxygen on Post Harvest Responses of Fruit

Reduced Respiration Rate

All living tissues respire taking in oxygen and giving out carbon dioxide. When the availability of oxygen is reduced, the rate of respiration also gets reduced.

Reduced Substrate Oxidation

During respiration, the complex organic molecules like carbohydrates, proteins etc, are broken down to simpler substances. Reduced respiration also reduces the substrate break down.

Delayed Ripening of Climacteric Fruit

During respiration there is a climacteric peak where the absorption of oxygen and release of carbon dioxide is high. Climacteric peak coincides with ripening and ripening is the initial stage of senescence. When oxygen supply is limited, climacteric peak is delayed and thereby ripening is also delayed.

Prolonged Storage Life

Reduced respiration rate, substrate oxidation and delayed ripening prolongs storage life of the commodity.

Delayed Breakdown of Chlorophyll

Oxygen is needed for chlorophyll breakdown. When oxygen availability is less, breakdown of chlorophyll is delayed and thereby the commodity retains its green color for a longer time.

Reduced Rate of Production of Ethylene

During ripening ethylene is produced. When ripening is delayed, rate of production of ethylene is also reduced.

Reduced Degradation Rate of Soluble Pectins

Pectins are responsible for firmness of the fruits. During ripening pectins are converted to soluble pectins and protopectins. When ripening is delayed due to low oxygen supply, pectin degradation is also delayed and fruit retains its firmness.

Effect of Increased CO_2 Levels

1. Decreased synthetic reactions in climacteric fruit
2. Delaying the initiation of ripening
3. Inhibition of some enzymatic reactions
4. Decreased production of some organic reactions
5. Reducing the rate of breakdown of pectic substances
6. Inhibition of chlorophyll breakdown
7. Retarded fungal growth on the crop
8. Inhibition of the effect of ethylene
9. Retention of tenderness
10. Decreased discolouration levels.

There are interacting effect of CO_2 and O_2 in extending the storage life of a crop and can be increased when they are combined.

Factors Deciding Optimum Composition of Gas

The optimum composition of gas varies for different produce and it depends on

1. Crop species
2. Crop cultivar
3. Physiological age of the crop at harvest
4. Degree of ripeness of the climacteric fruit
5. Growing conditions of the crop before harvest
6. Temperature
7. Concentration of gases in the store
8. Presence of ethylene in the store
9. Duration of treatment.

Negative Effects of CAS/MAS

Low levels of CO_2 above the optimum tolerable range can cause

1. Inhibition or aggravation of certain physiological disorders
2. Irregular ripening
3. Increased susceptibility to decay
4. Development of off-flavours
5. Loss of product quality.

Factors to be Considered with Controlled Atmosphere Storage

1. As it is expensive, only best fruits should be stored
2. Small fruits should be stored as they store better than large fruit

3. Fruits should be stored as soon as possible after harvest
4. The store should always be closed when not in use and cooled during loading
5. Only one type of crop should be stored in the same room and only one cultivar.

Table 15: Optimum Levels of O_2 and CO_2 for Long Term Storage of Some Tropical Fruits

Fruits	*Minimum Conc. of Tolerance Limit in % of O_2*
Avocado, banana, mangosteen, papaya, pineapple	2.0
Durian, mango, rambutan	3.0
Cherimoya, litchi, sapota, custard apple	5.0
Fruits	***Maximum Conc. of Tolerance Limit in % of Co_2***
Banana, mango	5.0
Avocado, Cherimoya, mangosteen, papaya, pineapple, sapota, custard apple	10.0
Rambutan	12.0
Durian, litchi	20.0

Controlled Atmosphere Technology

Store Structure

CA stores consist of insulated walls, ceiling and floor and these are made gastight. The insulation is covered with galvanized steel sheets which are fixed to the insulation. The joints between the sheets and between the walls and ceiling are coated with plastic or rubber to make them gastight. At the base of the walls the metal sheets let into the floor are also coated with plastic or rubber.

Table 16: Recommended CA Storage Conditions for Some Tropical Fruits

Fruits	*Temp. (°C)*	*RH (%)*	*Storage Life Weeks*	*CA*	
				% O_2	*% CO_2*
Avocado	7-13	90-95	2-4	2-5	3-10
Banana	13-14	90-95	1-4	2-5	2-5
Mango	10-14	85-90	1-4	2-5	5-10
Papaya	7-13	85-90	1-3	2-5	5-8
Pineapple	7-13	85-90	2-4	2-5	5-10

Modern stores have polyurethane or polystyrene foam sandwiched between two aluminum sheets. Gastight paint is also available which can be applied to the inner walls and ceiling of stores.

A major source of air leak in CAS is the door and on the door joint. When the door is closed the rubber gaskets press against each other to form a seal and insulated to ensure that the room is sufficiently airtight. A manometer is attached to it. Air is then evacuated with a pump to produce a vacuum. The manometer is then monitored to determine how long the vacuum is held.

Gas Control Equipment

The level of O_2 is constantly monitored by an oxygen analyzer and CO_2 by an infrared analyzer. This is normally set within acceptable tolerance limits of ±0.1 per cent. This means that for CO_2 where the required level is 5 per cent, the O_2 removal system would be activated when the level reaches 5.1 per cent and switched off when it reaches 4.9 per cent. For O_2 set at 1 per cent, the store would be ventilated when the O_2 level reaches to 0.9 per cent and switched off when it reaches 1.1 per cent.

Carbon Dioxide

CO_2 is controlled by removing it from the atmosphere by 'scrubbing'. If a low level (about 1 per cent) is required it is usually achieved by placing bags of hydrated lime [Ca $(OH)_2$] inside the store. Ca $(OH)_2$ reacts irreversibly with CO_2 and produce $CaCO_3$, water and heat. If a precise level of CO_2 is required the bags of lime can be placed in an adjacent room. When the CO_2 level is more than which is required, a fan draws the store atmosphere through the room containing the bags of lime until the required level is reached.About 25kg is required for each tonne of fruits for 6 months storage. After it is used in the store, the lime can be used for crop production purposes thus making the method economical.

Renewable scrubbers are also used to remove excess CO_2. These are compact and are suitable for use in CA transport systems. It consists of 2 containers with materials which can absorb CO_2. Air from the store is passed through one of these when it is required to reduce the CO_2 in the atmosphere. Activated charcoal or a molecular sieve (aluminum calcium silicate) are commonly used for this purpose. When the first container is saturated with CO_2 a device switches the air from the store to the second container. The first container then has fresh air blown through it to detach the CO_2 so that it can be used as a scrubber again when the second container is saturated. When a molecular sieve is used it is necessary to heat it during purging cycle. Activated charcoal can be purged simply by fresh air.

Oxygen

Levels of O_2 are controlled by ventilation. When levels are below than that is required a solenoid valve is activated by analyzer. This opens a ventilator to the outside and an electric fan is switched on to introduce fresh air into the store until the atmosphere has returned to the required O2 content. The ventilator should be opened before the cooling coils of the store to ensure an even temperature.

The speed at which optimum gas content of the store achieved has a significant effect on the quality of the fruit after storage.

Traditionally, high CO_2 and low O_2 is achieved by allowing the levels to be evolved by the metabolism of the product. This is called product generated CA. This can enhance the storage life of the crop. So it is common to fill the store with the crop, seal the store and inject nitrogen gas until the oxygen has reached the required level. The nitrogen gas can be obtained from N_2 cylinders.

Modified Atmosphere Storage (MAS)

It is defined as an atmosphere of the composition created by respiration or mixed and flushed into the product enclosure. This mixture is maintained through out the storage life and no further measurement or control takes place.

In modified atmosphere packaging or modified atmosphere storage, the fruit or vegetable is enclosed in sealed plastic film, which is slowly permeable to the respiratory gases. Within the package, the gases will change thus producing lower concentrations of O_2 and higher concentration of CO_2 than that in the fresh air

MAS is the development of a modified atmosphere around the product through the use of permeable polymeric films. It is an inexpensive method compared to controlled atmosphere storage. It is the alteration in the composition of gases in and around fresh produce by respiration and transpiration when such commodities are sealed in plastic films. These films restrict the accumulation and transmission of respiratory gases. This results in the accumulation of CO_2 and depletion of O_2 around the crop which increases the storage life.

The change in the atmosphere inside the sealed plastic film bag depends on the characteristics of the material used to make the package, the environment inside and outside the package as well as the respiration of the produce it contains. Changes in gas content may take sometime to reach equilibrium, but the speed at which the atmosphere is changed can be accelerated by gas flushing with N_2 to reduce the O_2 rapidly or the atmosphere can be flushed with an appropriate mixture of CO_2, O_2 and N_2 or the pack can be connected to a vacuum pump to remove the air so that the respiratory gases can change within the pack more quickly.

Table 17: Permeability of Different Packaging Films

Film	*Permeability (ml/m²/day at 1 atm)*		*$CO_2 : O_2$ Ratio*
	CO_2	*O_2*	
Polyethylene low density	7,700–77,000	3,900–13,000	2.0–5.9
Polyvinyl chloride	4,368–8,138	620–2,248	3.6–6.9
Poly propylene	7,700–21,000	1,300–6,400	3.3–5.9
Polystyrene	10,000–26,000	2,600–7,700	3.4–3.8
Saran	52–150	8–26	5.8–6.5
Polyester	180–390	52–130	3.0–3.5

Figure 49: Fruits and Vegetables with Modified Atmosphere Package

Modified Atmosphere Packaging (MAP)

Film or plastic materials that 'breathes' at a rate necessary to maintain correct mix of O_2, CO_2 and water vapour is used for packing.

1. When fresh produces is sealed inside a polymeric or plastic film package, respiration will lower the oxygen in the package and increases the CO_2 level.

2. In a well designed package with optimum permeability the gas levels within the pack will equilibrate in a range beneficial to the produce
3. This favourable unique atmosphere slows metabolic activities to a very low level.
4. Retains food and nutritional value
5. Increases shelf life and market flexibility
6. Good for branded, high quality, high value fruit and vegetables and minimally processed vegetables
7. Optimum CO_2/O_2concentrations are product specific and varies with cultivars /genotypes, production area, harvest maturity etc.

Advantages of High CO_2 and Low O_2 Atmosphere in MAP

1. Lowers respiration rate
2. Blocks biosynthesis of ethylene
3. Inhibits growth of pathogens
4. Maintains health and integrity of tissues
5. Prevents chlorophyll degradation
6. Maintains food value, nutritional value and flavour by slowing the loss of food reserves
7. Inhibit the loss of labile vitamins like Vitamin C and Vitamin A
8. Slows cell membrane degradation and loss of cellular compartmentalization and function.
9. Inhibit discolouration of cut surfaces

Absorbents in Modified Atmosphere Package

Ascorbic acid based sachets and catechol based sachets are used as O_2 absorbers. Calcium hydroxides acts as CO_2

absorbent. Potassium permanganate is used as ethylene absorbent. Commercial products available are Ethysorb and purafil.

Hypobaric Storage

It is another type of controlled atmosphere storage in which the storage chamber is placed under low pressure by application of vacuum. The reduction in pressure reduces the partial pressure of the O_2 and its availability to the crop. The reduction in the partial pressure of the O_2 is proportional to the reduction in pressure. As the crop in the hypobaric storage is constantly respiring, the store atmosphere should be constantly changed. This is achieved by a vacuum pump evacuating the air and the store atmosphere is constantly replenished from the outside. The air inlet and the air evacuation are balanced so as to achieve the required low pressure within the store.

Advantages

1. The O_2 level in the store can be accurately and easily achieved by simply measuring the pressure inside the store with a vacuum gauge.
2. Constant removal of ethylene gas from the store is also achieved.

The problems encountered in this method are

1. Low pressure may result in imploding
2. Reduced pressure inside the store result in rapid water loss from the crop.

Low Cost Storage

Refrigerated storage, which is the best method for storing fruits and vegetables in their fresh form is not only energy intensive but also involves huge capital investment. A low cost option to maintain lower temperature in an enclosed chamber has been a matter of prime importance. Evaporative cool storage or zero energy cool chamber is one such alternative system that is being explored in the country

Zero energy cool chambers were made of cheap quality porous bricks and riverbed sand. In the evaporative cooling a part or all of the sensible heat of moist air is converted to latent heat, thereby producing a reduction in temperature.

Under zero-energy cool chamber, the fruits can be stored up to 10 days with acceptable minimum rotting and quality loss, as against 5 days at room temperature.

It is an alternative to refrigerated cold storage to meet the short term storage needs at farm level especially during summer months when the temperature is very high and RH is relatively low, heavy weight loss due to transpiration from the freshly harvested produce is high. These structures at field level will maintain high RH (80-90 per cent) with temperatures close to the wet bulb, and therefore, reduces the weight losses drastically compared to those stored at ambient conditions.

Figure 50: Zero Energy or Evaporative Cool Chamber Made of Bricks and Sand

The advantages of zero energy or evaporative cool chamber

1. Low cost and eco-friendly
2. On farm storage
3. Enough time to decide the marketing destination of the produce.

Smaller version of a chamber at the field can be constructed using country bricks having double wall and the cavity is filled with river sand having 4 ducts submerged in the wet sand at the bottom to drain off the excess water from the river sand. Floor can be made of wooden planks with provision for entry of air. Temperature from 10 to 17°C less than the outside maximum can be obtained on a sunny day with lower RH outside, and sand fully saturated with water.

References

1. Abdullah, H., Rohaya, M.A., Zaipun, M.Z. 1985. Effect of modified atmosphere on black heart development and ascorbic acid contents in 'Mauritius' pineapple (*Ananas comosus cv.* Mauritius) during storage at low temperature. *Asean Food J.* 1: 15-18.
2. Aravind Prasad. B 1993. Refrigerated cold storage of fresh fruits and vegetables. Paper presented in the Advanced Technology Training Programme. 27th September 1993 at CFTRI Mysore.
3. Barkai-Golan, R. and Phillips, D. J. 1991. Post harvest heat treatment of fresh fruits and vegetables for decay control. *Plant Disease* 75: 1085-1089.

4. Bishop, D. 1996. Controlled atmosphere storage. In Dellino, C.J.V. (ed.) *Cold and Chilled Storage Technology,* Blacker, London.

5. Burdon, J.N. 1997. Post harvest handling of tropical and subtropical fruits for export. In *Post harvest physiology and storage of tropical and subtropical fruits* (ed.) Mitra, S.K., CAB Int. UK. pp.10.

6. Emerald, M.E.F., Sreenarayanan, V.V. 1999. Prolonging storage life of banana fruits by sub atmospheric pressure. *Indian Food Packer,* 49(3): 22-25.

7. Kader, A. 1980. Prevention of ripening of fruits by use of controlled atmosphere. *Food Technol.3:* 51-54.

8. Kader, A.A., Zagony, D., Kerbel, E.L. 1989. Modified atmosphere packaging of fruits and vegetables. CRC. *Rev. Food Sci. Nutr.* 28(1):1-30.

9. Kader, A.A. 1986. Biochemical and physiological basis for effects of controlled and modified atmospheres on fruits and vegetables. *Food Technol.,* 40(5): 100, 102,104

10. Kumar, T. and Kumar, D., 2004. Design of cold storage for fruits and vegetables. *Beverage and Food World,* 31: 20-22.

11. Nair, H., Tung, H.F. 1992. Low oxygen effect and storage of bananas. *Acta Hort.,* 292: 209-215.

12. Rao, G. 1989. Studies on storage of vegetables – Tomato. *Annual. Rep. Indian Inst. Hort. Res.,* Bangalore (India)

13. Resnizky, D., Sive, A. 1989. CA storage trials of mangoes Hanotea, 44(1): 53-56.

14. Sam, B.J. 1997. Shelf life of tomato. M.Sc. Hort. thesis, College of Horticulture, Kerala Agricultural University, 180p

15. Shashikumar Jain and Mukherjee, S. 2001. Zero energy cool chamber studies on mango. *Indian Food Packer,* 55: 55-58.

Chapter 8

Transportation

Distribution Systems and Hazards

Damage in transit is one of the oldest problems in packaging. The complete eradication of damage is not a realistic goal, because every hazard including accidents that packages meet cannot be anticipated. Protection is required against the average hazard encountered and not against the most severe that might occur in any particular journey. In practice, the absence of damage over a long period of time to any specific product visually indicates excessive packaging. Therefore, the main purpose of packaging is to provide produce with the attributes necessary to survive a number of different hazards which can be expected during storage, transportation and distribution.

The first step in selection of a package for a specific product and for a particular target market or markets is to form a clear picture of the distribution pattern which the product must follow. To do this, a distribution model should be drawn up. The distribution model is a representation of the system through which the produce is transferred, in qualitative as well as quantitative terms.

Distribution Patterns

The method of transportation is one of the most important factors to be taken into consideration in the choice of packaging to be utilized. Transport costs are considerable in the fruit and vegetable trade, and consequently packaging should be chosen to minimize these costs. Each distribution mode *viz.* road transport, rail transport, air transport and ship transport has its own particular characteristics with respect to the available technology constraints on package dimension and stresses imparted on the package goods. Conflicting demands are frequent due to the differences in these demands. Package designed for a particular distribution pattern will not be suitable for other distribution patterns without some modifications.

Road Transport

The principal hazards in road transport are vibration (repetitive shock) and bouncing of the package due to resonance. The vibration here is a function of loading and the suspension system characteristics of the vehicle concerned and also the road conditions. The vertical vibration causes more damage to the produce than the lateral and longitudinal. In order to prevent damage due to vibration, high head space in the package should be avoided. Commodities placed in the open trays should be equipped with net, film or paper covers secured to the tray. Some fruits are very sensitive to orientation, hence it is particularly important that the correct orientation is maintained during road transport.

Figure 51: Refrigerated Van Carrying Fruits and Vegetables

Even though stacking height (approx. 2m) is moderate in road transport, because of poorly maintained vehicles and roads, vibration results in the reduction of the stacking strength of the boxes. Also the effect of moisture on fibre board boxes further reduces the stacking strength of the boxes. Insulated vehicles are not adequate to prevent heating generated during respiration by fresh produce.

Temperature controlled vehicles (Refrigerated trucks) are desirable. In places where temperature controlled vehicles are not available, it is desirable to transport the produce during night time. For most of the produces, very high humidity (85-95 per cent) must be maintained to prevent transpiration losses. Normally there will be no feasibility in the vehicles to produce high humidity and maintain it. But, since the produce gives off moisture, humidity within the vehicle will be normally very high. In case of fiberboard containers, presence of high humidity will drastically reduces (nearly 50 per cent) resistance of the box to compressive force, since it absorbs moisture from the surrounding.

Air Transport

Compared to all other modes of transport, air transport is less hazardous in nature. Any damage to the produce in air transport is due to the associated ground operations and rough handling. The major hazard during transportation is relatively high frequency vibration from the air-craft engines and relatively low temperatures and pressures associated with flying at relatively high altitudes in unheated and unpressurised cabins. In air-crafts no cooling facility is available. The pre cooled product will keep sufficiently long until it can bc cooled at the destination point. This is true as long as there is no delay in starting and landing of air-crafts. The problem will be serious, if there is delayed transportation, because there is no proper facilities in airports for cooling and holding. This affects the quality of the produce due to heat produced during respiration. The packages in aircraft can be loose handled on pallets or in containers. The size of consignment varies from a full load to a few packages. Consignments of mixed

Figure 52: Air Transport

commodities are very common which will cause difficulties in handling and sorting operations.

Sea Transport

The hazards of sea transport are strongly dependent on storage conditions. The produce is normally stacked to 6 to 10m height and subjected to low frequency vibrations from the engine and propellers. In addition, pitching and rolling of the vessel can result in appreciable stresses, particularly in the lower levels of the Cargo. Regarding environmental conditions in ship transport, temperature and air refreshment control is very much required due to the length of the time involved. The produce will be pre cooled to a desired temperature before loading into the ship.

The refrigerated containers will only maintain the pre cooled temperature. Ventilation has to be sufficient in the container, in order to prevent accumulation of carbon dioxide and ethylene. For some produce, it is very essential to maintain very high humidity in the container to prevent weight loss and shrivelling of the produce due to moisture

Figure 53: Cargo for Shipment

loss. Mechanical humidity control is normally not possible in ship transport. Hence it is essential to prevent the moisture loss by some packaging means.

In case of corrugated fibre board boxes, the risk of the box during stacking is very high due to very high humidity surrounding the package and also the length of time involved in transportation. This problem can be overcome by using some water resistant coatings on the corrugated fibre board boxes or between surfaces of the outer liner. High air humidity increases the chance of rusting the stapler, nails, strapping seals and other steel components. For packaging of fresh produce, galvanized components will solve this problem.

Rail Transportation

Greatest distance exists between the production area and the main markets in India. Much of the produce has to travel under high temperature conditions, therefore the rail compartments meant for transporting such commodities needs extra attention in providing enough ventilations; such wagons can be provided with loovers for proper aeration, if not, refrigerated wagons must be preferred. But in India unlike other countries such wagons are not so common; we can also go for placing dry ice blocks wrapped in double heavy paper to facilitate handling and retard evaporation. They are placed below the roof in the centre of rail wagon so that the temperature can be brought down to18-20°C. Other hazards that can be encountered are vibration and impact, compression of improper stacking, development of high humidity and temperature, undue delay in transportation, jerks and oscillations due to rail joints, and the attack of rodents and insects. Hence adequate precautions has to be taken care by packers of fruits and vegetables intended to transport through rail.

Minimization of Transport Losses

Inadequate transport conditions, unsuitable containers and improper handling and stacking are some factors which contribute to serious product damage during transport. The losses are estimated to be millions of rupees worth. One vital step to alleviate this problem is to reduce the wastage by taking into consideration of the following factors.

Table 18

Sl.No.	Factors	Recommendations
1.	Long distance transport	Use rigid containers to protect commodities better. They do not break easily and protect the produce from injuries.
2.	Heat build up and gas concentration	Arrange the containers to allow air circulation freely on pallets or in containers. This prevents overheating of commodities and allow free exchange of gas between the commodity and the environment.
3.	Use of traditional containers such as bamboo baskets etc.	Use news paper / dried leaves or paddy straw to line the sides and bottom to protect the commodity from injuries due to sharp edges.
4.	Under packed and over packed commodities	Ensure that commodities are not under packed or over packed. Under packed containers result in more vibration damage while over packed containers increase damage due to compression.
5.	Control of respiration and transpiration	Transport the commodities preferably at dawn or night when temperature is low. This inhibits fast respiration and water loss from the products.
6.	Up and down movement of containers inside the vehicles	Use vehicles with good shock absorbers

Contd...

Table 18–Contd...

Sl.No.	Factors	Recommendations
7.	Space for ventilation in the vehicle	Allow a space between the cover/ top of truck and the produce for ventilation
8.	To reflect the heat of sun during day transport	Paint the canvas/roof of the truck with white color
9.	Use of deep containers	Avoid using deep containers since the produce at the bottom will be more compressed and causes injuries
10.	Placement of fragile materials inside the vehicle	Load more fragile materials near the centre of vehicles since the vibrational inputs are less compared to front and rear ends of the vehicle.

Hazards in Manual Handling

The hazards of manual handling of individual packages are serious in nature. In case of palletized and containerized packages, though the problem is reduced but not completely eliminated. Manual handling is very much essential in case of trans shipment and for transfers from one pallet to another and from one storage shelf to another. Many operations are carried out by persons who lack knowledge about the fragility of fresh produce and have no incentive to handle gently.

Minimization of Handling Losses

Following are some of the measures which can help to minimize dropping hazards.

1. International fragility symbol on all packages carrying fragile produce
2. International orientation symbol with the words "THIS WAY UP"

3. Providing the packages with handles or hand holes for easy handling.
4. Avoiding the use of very light packs.

References

1. Fallik, E., Aharoni, Y., Yekutieli, O., Wiseblum, A., Regev, R. Beres, H. and Bar Lev, E. 1996. A method for simultaneous cleaning and disinfecting agricultural produce. Israel Patent Application No. 116965.
2. Ladaniya, M. S. and Shankar, R.K. 1996. Containers for packaging and transportation of Nagpur mandarin. Annual report, 1995-96,National Research Centre on Citrus, Nagpur, 57-5.
3. McGregor, B. 1987. Tropical products transport handbook. USDA, Office of transportation, Agricultural handbook. 668.
4. Roy, S. K. and Pal, K. 2000. Latest techniques on transportation and storage of horticultural produce. Souvenir of the National Seminar on Hi-tech Horticulture. 26-28 Jan. 2000. Bangalore. p 93.
5. Ryall, A.Land Lipton, W.J. 1972. Handling, transportation and storage of fruits and vegetables. The AVI Publishing Company, Inc. p. 462.
6. Sathish, H.S. 1993. Distribution hazards and packaging of fresh produce. Paper presented in the Advanced Technology Training Programme, 27 September 1993, CFTRI.
7. Yahia, M.E. 1999. Modified and Controlled atmospheres for tropical fruits. Hort. Rev. 22: 123-183.

Index

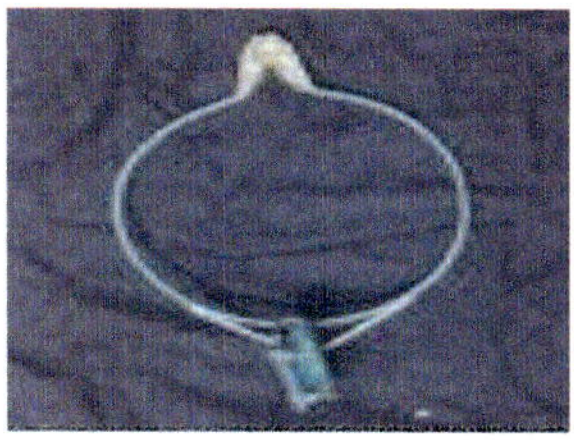

Figure 2: Mango Harvester (Page No. 12)

Figure 3: Fruits and Vegetables for Washing is Taken in the Inner Vessel having Full of Perforations and Repeatedly Immersed in the Outer Vessel Containing Water with 30 ppm Chlorine. Quantity taken in the inner vessel is that sufficient enough to be handled by a person. (Page No. 27)

Figure 4: Soak Tank (Page No. 29)

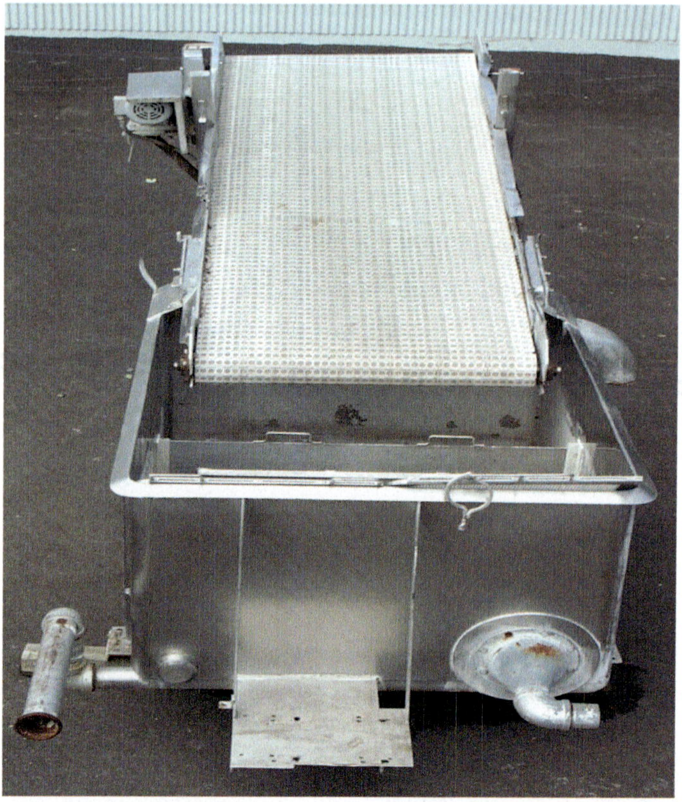

Figure 5: Soak Tank, Conveyor Belt with Perforations (Page No. 29)

Figure 6: Rotary Barrel Washer (Page No. 30)

Figure 7: Rotary Barrel with Holes (Page No. 30)

Figure 8: Rotary Barrel with Agitator (Page No. 31)

Figure 9: Spray Washer (Page No. 32)

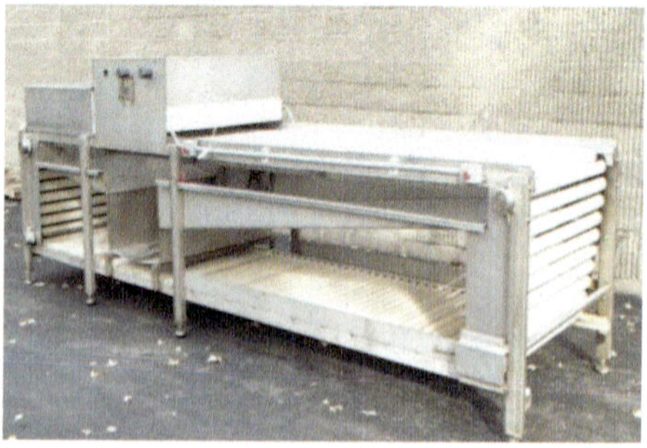

Figure 10: Spray Washer cum Inspection Table (Page No. 32)

Figure 11: Brush Washer (Page No. 33)

Big – 8.5 –7.5 cm

Medium –7.49-6.5 cm

Small – 6.49 – 5.5 cm

Figure 12 (Page No. 40)

Poor– <130g

Average –131-170 g

Good – 171–200 g

Excellent– >200 g

Figure 13: Excellent (Page No. 41)

Small– <1 kg

Medium–1 –2 kg

Large–>2 kg

Figure 14: Large (Page No. 41)

A class–1500-1800 g

B class–1100-1500 g

C class–900-1100 g

D class–700-900 g

Figure 15: Class A (Page No. 41)

Small–< 25 g

Medium–25-50 g

Large–50-75 g

Extra large–> 75 g

FIgure 16: Medium Size (Page No. 42)

Figure 17: Weight Graders (Page No. 50)

Figure 18: IIHR Electronic Weight Grader
1: Grading unit; 2: Singulation unit; 3: Control unit; 4: Weighing unit (Page No. 51)

Figure 19: Singulation Unit (Page No. 51)

Figure 20(a): Screen Grader (Page No. 54)

Figure 20(b): Screen Magnified (Page No. 54)

Figure 21: IARI Potato Grader (Page No. 55)

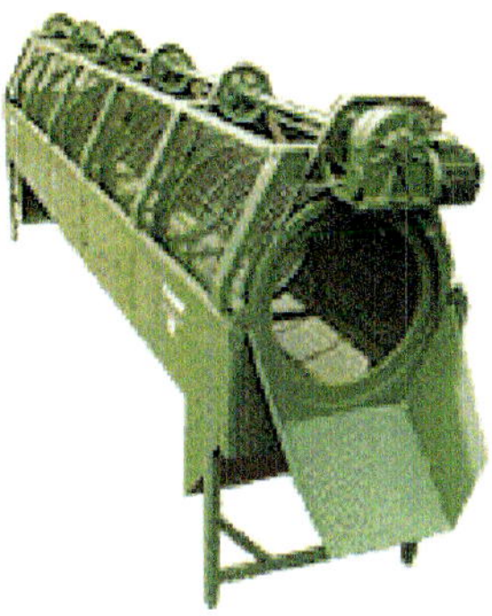

Figure 22: Rotary Barrel/Drum Type Grader (Page No. 56)

Figure 23: Divergent Opening (Roller) Two Stage (Page No. 57)

Figure 24: Divergent Opening (Roller) Single Stage (Page No. 57)

Figure 26: Expanding Opening Single Belt Grader (CIPHET, Ludhiana) (Page No. 59)

Figure 27: Expanding Opening Roller Grader (Page No. 60)

Figure 28: Onion Grading (Page No. 61)

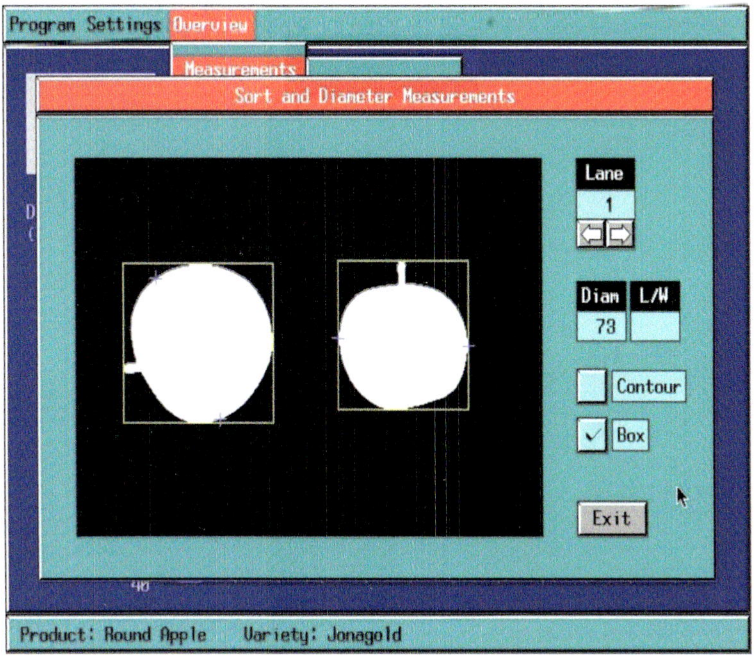

Figure 29: Size Sensing (Page No. 62)

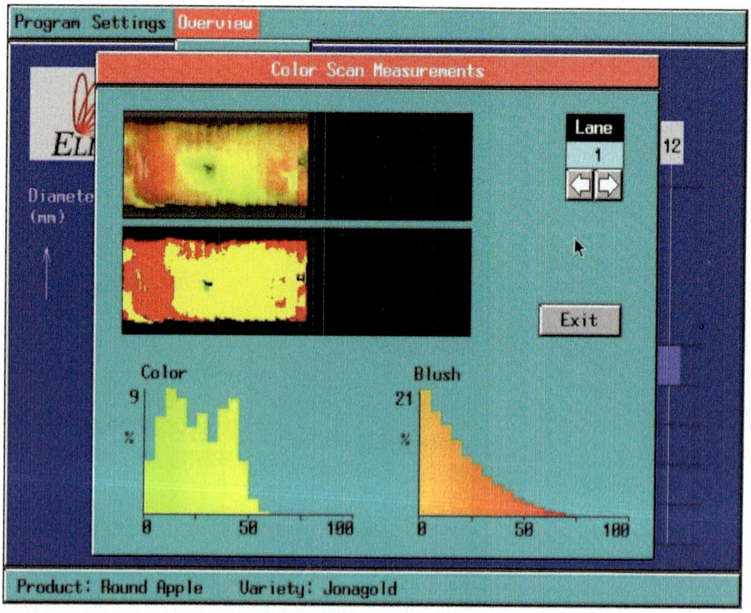

Figure 30: Colour Sensing through Cameras (Page No. 63)

Figure 31: Water Cooling Chamber
Vegetables immersed in ice water for pre-cooling
(Page No. 75)

Figure 32: Forced Air Cooling Unit
Fruits packed in ventilated cartons for pre-cooling fast method but more expensive (Page No. 76)

Figure 33: Bamboo Baskets with Mangoes (Page No. 88)

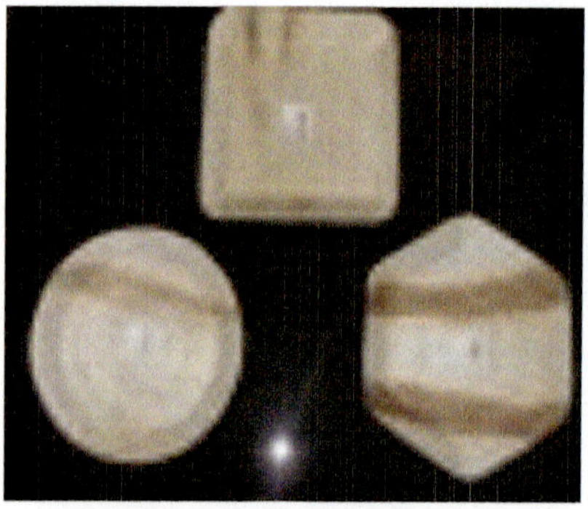

Figure 34: Arecanut Leaves Sheath Trays (Page No. 88)

Figure 35: Jute Bags with Grains (Page No. 89)

Figure 36: Wooden Boxes with Tomatoes (Page No. 90)

Figure 38: Corrugated Fibre Board Box with Vegetables (Page No. 92)

Figure 39: Waxed Chip Board Cartons with Fruits (Page No. 92)

Figure 40: Plastic Crates with Potatoes (Page No. 93)

Figure 41: Moulded Trays with Apple (Page No. 94)

Figure 42: Thermoformed Plastic Trays with Fruits (Page No. 95)

Figure 43: Plastic Bag Containing Apples (Page No. 96)

Figure 44: Plastic Box with Okra (Page No. 97)

Figure 45: Carrot Individually Wrapped with Cling Film (Page No. 98)

Figure 46: Shrink Wrapping Machine (Page No. 99)

Figure 47: Active Packaged Banana Hands with Sachets Placed in the Centre Containing the Active Agents (Page No. 101)

Figure 48: Fruits and Vegetables in Cold Room (Page No. 107)

Figure 49: Fruits and Vegetables with
Modified Atmosphere Package (Page No. 128)

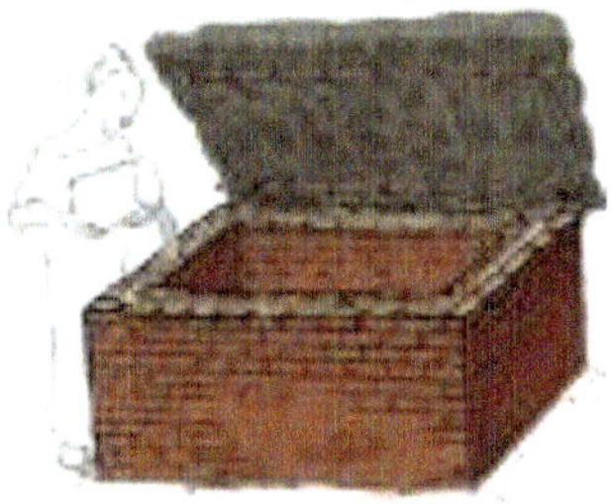

Figure 50: Zero Energy or Evaporative Cool Chamber Made of Bricks and Sand (Page No. 131)

Figure 51: Refrigerated Van Carrying Fruits and Vegetables (Page No. 137)

Figure 52: Air Transport (Page No. 138)

Figure 53: Cargo for Shipment (Page No. 139)